Lecture Notes in Control and Information Sciences

Edited by M. Thoma and A. Wyner

Vol. 117: K.J. Hunt
Stochastic Optimal Control Theory
with Application in Self-Tuning Control
X, 308 pages, 1989.

Vol. 118: L. Dai
Singular Control Systems
IX, 332 pages, 1989

Vol. 119: T. Başar, P. Bernhard
Differential Games and Applications
VII, 201 pages, 1989

Vol. 120: L. Trave, A. Titli, A. M. Tarras
Large Scale Systems:
Decentralization, Structure Constraints
and Fixed Modes
XIV, 384 pages, 1989

Vol. 121: A. Blaquière (Editor)
Modeling and Control of Systems
in Engineering, Quantum Mechanics,
Economics and Biosciences
Proceedings of the Bellman Continuum
Workshop 1988, June 13–14, Sophia Antipolis, France
XXVI, 519 pages, 1989

Vol. 122: J. Descusse, M. Fliess, A. Isidori,
D. Leborgne (Eds.)
New Trends in Nonlinear Control Theory
Proceedings of an International
Conference on Nonlinear Systems,
Nantes, France, June 13–17, 1988
VIII, 528 pages, 1989

Vol. 123: C. W. de Silva, A. G. J. MacFarlane
Knowledge-Based Control with
Application to Robots
X, 196 pages, 1989

Vol. 124: A. A. Bahnasawi, M. S. Mahmoud
Control of Partially-Known Dynamical Systems
XI, 228 pages, 1989

Vol. 125: J. Simon (Ed.)
Control of Boundaries and Stabilization
Proceedings of the IFIP WG 7.2 Conference
Clermont Ferrand, France, June 20–23, 1988
IX, 266 pages, 1989

Vol. 126: N. Christopeit, K. Helmes
M. Kohlmann (Eds.)
Stochastic Differential Systems
Proceedings of the 4th Bad Honnef Conference
June 20–24, 1988
IX, 342 pages, 1989

Vol.127: C. Heij
Deterministic Identification
of Dynamical Systems
VI, 292 pages, 1989

Vol. 128: G. Einarsson, T. Ericson,
I. Ingemarsson, R. Johannesson,
K. Zigangirov, C.-E. Sundberg
Topics in Coding Theory
VII, 176 pages, 1989

Vol. 129: W. A.Porter, S. C. Kak (Eds.)
Advances in Communications and
Signal Processing
VI, 376 pages, 1989.

Vol. 130: W. A. Porter, S. C. Kak, J. L. Aravena (Eds.)
Advances in Computing and Control
VI, 367 pages, 1989

Vol. 131: S. M. Joshi
Control of Large Flexible Space Structures
IX, 196 pages, 1989.

Vol. 132: W.-Y. Ng
Interactive Multi-Objective Programming
as a Framework for Computer-Aided Control
System Design
XV, 182 pages, 1989.

Vol. 133: R. P. Leland
Stochastic Models for Laser Propagation
in Atmospheric Turbulence
VII, 145 pages, 1989.

Vol. 134: X. J. Zhang
Auxiliary Signal Design in Fault
Detection and Diagnosis
XII, 213 pages, 1989

Vol. 135: H. Nijmeijer, J. M. Schumacher (Eds.)
Three Decades of Mathematical System Theory
A Collection of Surveys at the Occasion of the
50th Birthday of Jan C. Willems
VI, 562 pages, 1989

Vol. 136: J. Zabczyk (Ed.)
Stochastic Systems and Optimization
Proceedings of the 6th IFIP WG 7.1
Working Conference,
Warsaw, Poland, September 12–16, 1988
VI, 374 pages. 1989

For information about Vols. 1–116 please contact your bookseller or Springer-Verlag

Lecture Notes
in Control and Information Sciences

182

Editors: M. Thoma and W. Wyner

J. Hagenauer (Ed.)

Advanced Methods for Satellite and Deep Space Communications

Proceedings of an International Seminar
Organized by Deutsche Forschungsanstalt
für Luft- und Raumfahrt (DLR)
Bonn, Germany, September 1992

Springer-Verlag Berlin Heidelberg GmbH

Advisory Board

Editor

Prof. Dr. Joachim Hagenauer
Institut für Nachrichtentechnik
Deutsche Forschungsanstalt
für Luft- und Raumfahrt (DLR)
W-8031 Oberpfaffenhofen
Germany

ISBN 978-3-540-55851-4   ISBN 978-3-540-47299-5 (eBook)
DOI 10.1007/978-3-540-47299-5

Typesetting: Camera ready by authors

61/3020 5 4 3 2 1 0 Printed on acid-free paper

Foreword

We are currently observing a discussion about the usefulness of space resarch and the amount of spin-off we obtain from space programs. One area which is widely overlooked is digital communications, although it is quite clear that the application of space research which has directly affected most people is certainly space communication via satellite. Direct broadcasting satellites, international phone calls via geostationary satellites and the growing mobile services by satellites serve more and more users.

Digital modulation and coding which were theoretically developed following Shannon's famous paper in 1948, found their first applications in space research. PSK modulation, phase-locked loops, and Costas loops, convolutional codes, concatenated codes, Reed-Solomon-Codes, sequential decoding and Viterbi decoding were for the first time applied in satellite and deep-space communications. Funding in the early days mostly came from space programs of NASA, ESA, INTELSAT and national ministries. Nowaday all those techniques are used in consumer electronics, like CD-players, FM-receivers, digital audio broadcasting, cellular phone systems, data modems and the upcoming digital TV and HDTV-receivers. The spin-off from these techniques funded by space-research programs is not only in technology like VLSI-implementation but also in theory, in new algorithms and in new system design.

Some 10 to 5 years ago many people believed that satellite communications would soon be outdated by high data rate optical fiber communications. Today we experience another boom in satellite communications in many areas where the flexibility and wide range availability of satellites are important. Examples are the use of very small aperture terminals (VSAT) in business communications, broadcast satellites and mobile services for communication and navigation for ships, airplanes, trucks, private cars and personal world wide communications services for the modern traveller. And even the long-haul trunk services of INTELSAT experience an increasing demand, because of their flexibility to set up links quickly and to any location when needed.

Despite the fact that satellite communications are widely and economically applied, further research is needed because more and more services compete for the limited resource of radio frequencies and cause increasing mutual interference. We need bandwidth efficient modulation schemes, high compression factors for source codes for voice, audio, image and TV transmission . Further research is needed into error correcting schemes for low error rates, combined source-, channel- and modulation codes and for systems which use switching in space and can utilize higher radio frequencies (20/30/60 GHz and above) as well as optical frequencies.

We all were excited about the fantastic images we obtained from space probes like Giotto and Voyager from comets and planets as distant as Jupiter and Pluto. This was not only a triumph for planetary research but also a world (or rather universe) record for communications. Signals were transmitted with a 20W transmitter over a distance of 7 million km. The receivers operated very close to the absolute limit set by Shannon's theory.

The "mission to planet earth" will use a large set of earth observation satellites, synthetic aperture radars and high resolution cameras to monitor our fragile environment. However, the images and data from low orbiting and geostationary observation platforms have to be transmitted to earth and the bottlenecks in transmission capacity are foreseeable.

During the 1992 DLR-Seminar

"Advanced Methods for Satellite and Deep Space Communications"

internationally well known scientists describe the state of the art, as well as new methods and systems for satellite and space communications. Therefore, the proceedings of this seminar should be a valuable source of information in the area of advanced communications for scientists, engineers and for decision making managers in government, service providers and industry.

Joachim Hagenauer
Oberpfaffenhofen

Contents

Deep Space Communications and Coding: A Marriage Made in Heaven   1
James L. Massey, ETH Zürich, Switzerland

Coding and Coded Modulation Techniques for Reliable Satellite and Space   18
Communications
Shu Lin, Sandeep Rajpal and Do Jun Rhee, University of Hawaii at Manoa, USA

Synchronization Aspects of a Mobile Satellite Voice and Data Modem   40
B. Koblents, P.J. McLane and W. Choy, Queen's University, Kingston, Canada

Advanced Modulation Formats for Satellite Communications   61
Ezio Biglieri, Politecnico di Torino, Italy

Generalizations of the Viterbi Algorithm with Applications in Radio Systems   81
Carl-Erik W. Sundberg, AT&T Bell Laboratories, Murray Hill, USA

Advanced Communications Technologies for Future INTELSAT Systems   100
Peter P. Nuspl, INTELSAT, Washington D.C., USA

Personal Handheld Communications via Hybrid K_a- and L/S-Band Satellites   122
Russell J. F. Fang, COMSAT Laboratories, Clarksburg, USA

Coherent Optical Space Communications   135
Walter R. Leeb, Technische Universität Wien, Austria

The NASA Program for Optical Deep-Space Communications at JPL   159
James R. Lesh, Jet Propulsion Laboratory, Pasadena, California, USA

DLR Experimental Systems for Free Space Optical Communications   180
J. Franz, Ch. Rapp and B. Wandernoth, DLR Oberpfaffenhofen, Germany

Deep-Space Communications and Coding: A Marriage Made in Heaven

James L. Massey
Signal and Information Processing Laboratory
Swiss Federal Institute of Technology
CH-8092 Zürich, Switzerland

1 Introduction

It is almost a quarter of a century since the launch in 1968 of NASA's Pioneer 9 spacecraft on the first mission into deep-space that relied on coding to enhance communications on the critical downlink channel. [The channel code used was a binary convolutional code that was decoded with sequential decoding--we will have much to say about this code in the sequel.] The success of this channel coding system had repercussions that extended far beyond NASA's space program. It is no exaggeration to say that the Pioneer 9 mission provided communications engineers with the first incontrovertible demonstration of the practical utility of channel coding techniques and thereby paved the way for the successful application of coding to many other channels.

Shannon, in his 1948 paper [1] that established the new field of information theory, gave a mathematical proof that every communications channel could be characterized by a single parameter C, the channel capacity, in the manner that information could be sent over this channel to a destination as reliably as desired at any rate $\mathcal{R}$ (measured, say, in information bits per second) provided that $\mathcal{R} < C$, but that for any rate $\mathcal{R}$ greater than C there was an irreducible unreliability for information transmission. Shannon's work showed that, to achieve efficient (i. e., $\mathcal{R}$ close to C) and reliable use of a channel, it was necessary in general to "code" the information for transmission over the channel in the sense that each transmitted digit must depend on many information bits. Almost immediately, there began an intensive search (that still continues unabated) by many investigators to find good channel codes. Unfortunately, most of these researchers concentrated (and still do concentrate) on the "wrong" channel, viz. on the binary symmetric channel (BSC) [or its non-binary equivalents]. The BSC has both a binary input alphabet and a binary output alphabet and is characterized by a single parameter p ($0 \leq p \leq 1/2$) in the manner that each transmitted binary digit has probability p of being changed in transmission, independently of what has happened to the previous transmitted digits (i. e., the channel is memoryless). It was natural then to think of such a channel as introducing "errors" into the transmitted data stream and to suppose that it was the purpose of the coding system to "correct" these errors. The term "error-correcting code" came into (and remains) in widespread use to describe such channel codes--although with scarcely any reflection one must conclude that one cannot even talk about "errors" in transmission unless the channel input and output alphabets coincide (as unhappily they do for the BSC). More circumspect writers have adopted the term "error-control code" in place of "error-correcting code," but this is only a trifle less misleading. Where are the errors to be controlled? It seems to us much wiser to use the less suggestive, but more precise, term "channel code" to describe the code used to map the information bits into the sequence of digits to be transmitted over the channel, as we have done already in our opening paragraph.

There had been attempts prior to 1968 to make practical use of channel coding. Codex Corporation, founded in 1962 in Cambridge, Mass., became the first company dedicated to this goal and also the first to encounter the widespread skepticism among communications engineers about the practicality of channel coding. The considerable commercial success that has been enjoyed by this company (which is now a division of Motorola, Inc.) stems less from its pioneering activity in channel coding than from its judicious decision in the late 1960's to expand its technical activity into the development of high-speed modems for telephone channels.

Why did deep-space communications provide the setting for demonstrating the practical benefits of channel coding? Why were the "heavens" of deep-space virtually predestined to be the proving grounds for channel coding techniques? Why was this wedding of channel coding to the deep-space channel, in the words of our title, a "marriage made in heaven"? There are many reasons, the most important of which are as follows:

• The deep-space channel is accurately described by a mathematical channel model, the additive white Gaussian noise (AWGN) channel, that was introduced by Shannon in 1948.

• It was well understood theoretically by the early 1960's what one must do to use this channel efficiently and reliably and what gains could be achieved by channel coding.

• The available bandwidth on the deep-space channel was so great that binary transmission could be efficiently used.

• The NASA communications engineers in the mid-1960's understood that, for efficient use of the deep-space channel, it was necessary to design the modulation system and the channel coding system in a coordinated way and they were willing to make the resulting necessary changes in the demodulators that they had previously been using.

• Good binary convolutional codes were available by the mid-1960's and, more importantly, an effective and practical algorithm was known for decoding these codes on the AWGN channel.

• The only complex part of the efficient channel coding systems for the AWGN channel that were known by the mid-1960's is the decoder, which for the downlink deep-space channel is located at the earth station where complexity is of much less importance than it is in the spacecraft.

• Every dB in "coding gain" on the downlink deep-space channel is so valuable (in the mid-1960's it was reckoned at about $1,000,000 per dB) that even a small gain, such as the 2.2 dB that was provided by the Mariner '69 channel coding system, was a strong economic incentive for developing and implementing such a channel coding system.

We will consider most of the above reasons in some detail in the sequel--there are lessons therein that are still of great relevance today and are all-too-often forgotten. We begin in the next section by taking a careful look at the deep-space channel itself, then considering the interplay between coding and modulation systems used on this channel. We also make a fairly intensive study of bandwidth issues for the deep-space channel that leads us to the important distinction between Fourier bandwidth and Shannon bandwidth. The final section is devoted to a detailed look at the two codes in question, the Pioneer 9 code and the Mariner '69 codes, followed by a comparison of their merits. It is something of a quirk in technical history that the Pioneer 9 convolutional coding system became the first channel coding system to be used in deep-space, as this had not been intended by NASA. That honor had been planned for the binary block code used in NASA's Mariner spacecraft that was launched in 1969. Why the convolutional code nevertheless won the race into space is explained in the final paragraph of this "tale of two codes."

2 The Deep-Space Channel

2.1 Channel Capacity Considerations

We have already mentioned that the deep-space channel is accurately described by Shannon's additive white Gaussian noise (AWGN) channel model. The correspondence is so good in fact that no one has ever observed any deviation of the deep-space channel from this mathematical model. In this model, the received signal is the sum of the transmitted signal and a white Gaussian noise process of one-sided power spectral density N_0 (watts/Hz).

The transmitted signal is constrained to lie in a Fourier bandwidth of W (Hz) or less and to have an average power of S (watts) or less. Shannon [1] computed the capacity of this channel to be

$$C_W = W \log_2 (1 + \frac{S}{W N_0}) \quad \text{(bits/sec)}, \tag{1}$$

which is one of the most famous formulas in communication theory--and also one of the most abused. It is obvious intuitively (increasing the available Fourier bandwidth can only help the sender) and easy to check mathematically that C_W increases monotonically with W, taking as its maximum the value

$$C_\infty = \frac{1}{ln\ 2} \frac{S}{N_0} \approx 1.44 \frac{S}{N_0} \quad \text{(bits/sec)}. \tag{2}$$

Suppose now that one is transmitting information bits at a rate $\mathcal{R}$ (bits/sec) very close to this maximum capacity. Then, because the power is S (joules/sec), the energy per information bit, E_b, is just $E_b = S/\mathcal{R} \approx S/C_\infty = ln\ 2\ N_0 \approx 0.69\ N_0$ (joules). Equivalently, the *signal-to-noise ratio* is

$$\frac{E_b}{N_0} \approx 0.69 \quad \text{(or -1.6 dB)}, \tag{3}$$

which is the minimum signal-to-noise ratio required for arbitrarily reliable communication and is often referred to as the *Shannon limit* for the AWGN channel. All of this was well known in the early 1960's, cf. [2, p. 162].

2.2 The Interplay between Coding and Modulation Systems

Suppose now that digital transmission is used on the AWGN channel and that the modulator emits transmitted symbols [or, more precisely, the waveforms that represent these symbols] at the rate of r (symbols/sec). Of course, $S = r \times E_b = \mathcal{R} \times E_b$ so that E_b and E are related as $E_b = E (r/\mathcal{R})$. Incidentally, one of the benefits derived from channel coding for the deep-space channel is that it accustomed perspicacious communications engineers to evaluate the performance of their communications systems in terms of the "true" signal-to-noise ratio defined as E_b/N_0, which is a fundamental performance parameter, rather than with the transmitted signal-to-noise ratio, E/N_0, where E (joules/symbol) is the average energy of the waveforms used for a transmitted symbol, which is not fundamental at all for comparison of system performances although it had been customarily so used (and is still so used unfortunately often).

Suppose next that there are q signals in the modulation *signal set*, i.e. in the set of waveforms used to reqresent the q different values of a modulation symbol. It is convenient to represent these signals or waveforms as vectors in n-dimensional Euclidean space in the manner introduced by Shannon [2] and exploited to great effect by Wozencraft and Jacobs [3], i. e., by their coefficient vectors with respect to their representation as a linear combination of the signals in some orthonormal set of n signals. Let $s_0, s_1, ..., s_{q-1}$ so represent the modulation signal set. The binary one-dimensional (or scalar) signal set $s_0 = +\sqrt{E}$ and $s_1 = -\sqrt{E}$ is called the *binary antipodal signal set* and represents, for instance, the waveforms used in binary phase-shit-keying (BPSK) modulation. BPSK modulation is very attractive for use in space communications because its constant-envelope character greatly simplifies the required transmitter amplifying hardware in the spacecraft.

The use of digital modulation on the AWGN channel effectively converts the channel to a discrete-time additive Gaussian noise channel in which the received signal r in each modulation interval is the sum of the transmitted signal s_i and a noise vector n whose components are independent Gaussian random variables with 0 mean and variance $N_0/2$. The capacity C (bits/use) of this channel was also well-known in the early 1960's. In particular, it was known that if one uses a one-dimensional signal set according to a probability distribution on the signals such that the received signal r well-approximates a zero-mean Gaussian random variable, then (cf. [4, p. 147]) the capacity C is given by

$$C \approx \frac{1}{2} \log_2 (1 + \frac{2E}{N_0}) \quad \text{(bits/use)}. \tag{4}$$

One sees from (4) that C/E, the capacity per joule, decreases as E/No increases. In the region of energy-efficient operation, which is roughly the region $0 < E/N_0 \le 1/2$ (-3 dB), the capacity (4) becomes

$$C = 1.44 \frac{E}{N_0} \quad \text{(bits/use)}. \tag{5}$$

The corresponding capacity per unit of time is thus

$$r C = 1.44 \frac{r E}{N_0} = 1.44 \frac{S}{N_0} = C_\infty \quad \text{(bits/sec)} \tag{6}$$

where we have made use of (2). Thus, one sees that in the region of energy-efficient operation, which is roughly the region $0 < E/N_0 \le 1/2$, one pays no penalty in capacity for using one-dimensional digital modulation. If the one-dimensional signal set is the binary antipodal signal set and the two signals therein are equally likely, then the received signal r = s + n always has zero mean but will have the required approximately Gaussian distribution only if the standard deviation $\sqrt{N_0/2}$ of the Gaussian noise n is somewhat greater than the magnitude $\sqrt{E}$ of s, i. e., roughly again when $0 < E/N_0 \leq 1/2$. The conclusion that *binary antipodal modulation is energy-efficient on the deep-space channel just when one-dimensional modulation is energy-efficient* i. e., when the transmitted signal-to-noise ratio E/N_0 is about -3 dB or less, was well-known to information theorists in the early 1960's.

Suppose that information bits are sent *uncoded* over the deep-space channel with binary antipodal modulation, which implies that the number of modulation symbols per second, r, is equal to the number of information bits per second, $\mathcal{R}$ and hence that $E_b = E$. The probability of detecting the information bits at the receiver reduces to that of detecting equally likely binary antipodal signals, $+\sqrt{E_b}$ and $-\sqrt{E_b}$, in the presence of Gaussian noise with variance $N_0/2$, the error probability for which (cf. [3, p. 82]) is given by

$$P_b = Q(\sqrt{2E_b/N_0}) \tag{7}$$

where

$$Q(x) = \int_x^\infty \frac{1}{\sqrt{2\pi}} e^{-\alpha^2/2} \, d\alpha < \frac{1}{\sqrt{2\pi}\,x} e^{-x^2/2} \tag{8}$$

and where the inequality in (8) is a virtual equality for $x \geq 2$ (cf. [3, p. 83]). If one operates in the region of potentially efficient channel use, $E_b/N_0 = E/N_0 \leq 1/2$ (-3 dB), one sees from (7) and (8) that the information bits can be recovered at the receiver with an error probability of at best $P_b = Q(1) \approx 2.4 \times 10^{-1}$, which is orders-of-magnitude too large to be acceptable on the downlink deep-space channel or in almost any communications system for that matter. The inescapable conclusion that *if one wants to signal both energy-efficiently and reliably with binary modulation on the deep-space channel, then one must use channel coding.* Indeed, it was the realization of this fact in the early 1960's that impelled NASA to begin to plan for the use of channel coding in the spacecraft in its Mariner series that would be launched in 1969. With the long lead time required to obtain space approval for design changes, it was already too late to influence spacecraft in the Mariner series with earlier launch dates--these spacecraft continued to use uncoded binary antipodal signalling on the downlink.

The relationship of demodulation to the channel coding system is much more subtle than that of modulation. The reason for this is that the demodulator can spoil the channel for decoding if one is not extremely careful about its design, a fact that is still not sufficiently appreciated. Before channel coding theory "reared its ugly head", communications engineers designed their digital demodulators to make optimum decisions about the transmitted modulation symbols, i. e. they did what is now generally called *hard-decision demodulation*. The combination of modulator, waveform channel and hard-decision demodulator creates a discrete channel whose input and output alphabets coincide. In particular, when binary antipodal modulation is used on the deep channel with hard-decision demodulation, the resulting discrete channel is the binary symmetric channel (BSC) about which we have already made disparaging remarks; the "error probability" p on this channel is given by $p = Q(2E/N_0)$. [One sees here that the BSC does not occur naturally in nature; it is created by the designers of energy-inefficient modulation systems!] Nonetheless, most communications engineers continued long after 1948 to assume (and many today still do assume) that the proper goal of of a digital demodulator is to make optimum decisions about the transmitted modulation symbols, i. e., to do hard-decision demodulation. This slothful thinking meshed very nicely with the "error-correcting codes" school of channel coding theorists since hard-decisions give the "errors" that they were eager to correct. It was, however, well-known to many information theorists in the early 1960's that hard-decision demodulation entails a substantial loss in capacity compared to what can be achieved by a more thoughtful form of demodulation. For instance, it was well-known that binary antipodal signalling on the deep-space channel used with hard-decision demodulation achieves a capacity smaller than C_∞ by a factor of $2/\pi$ (2.0 dB) in the energy-efficient range of operation, cf. [4, p. 211]. Because the coding system for Mariner '69 was calculated to provide only 2.2 dB of gain over uncoded binary transmission even when used with optimum demodulation, the use of hard-decision demodulation in Mariner '69 was out of the question!

What should a good demodulator do? Fano answered this question very well in the early 1960's: "the capacity of the discrete channel [that results from the combination of the digital modulator, waveform channel and demodulator] should not be unduly smaller than the capacity of the original [waveform] channel" [4, p. 211]. The real goal of the designer of the demodulation system should be to create a good channel for the channel coding system--a point that we have written about at greater length elsewhere [5]. If the capacity of the resulting discrete channel is taken as the design criterion, then one must conclude that the optimum demodulator is a straight wire because any quantization of the received signal can only reduce capacity. In fact, to ease the decoding problem for the channel code, one should in fact design the demodulator to make as coarse a quantization of the received signal as possible consistent with Fano's adage that the capacity of the [resulting] discrete channel should not be unduly smaller than the capacity of the original [waveform] channel." Already in the early sixties it was well-known among some information theorists, cf. [6], that 8 demodulator quantization levels were enough, when used for binary antipodal signalling on the deep-space channel, to reduce the capacity loss to a negligible 0.1 dB in the energy efficient range $E/N_0 \leq 1/2$ (-3 dB). However, using only 4 quantization levels would yield an additional loss of about 0.3 dB. It was natural then to choose 8 level demodulation, or "3-bit soft-decision demodulation" as it is generally called because the soft-decision demodulator's output can be thought to consist of the hard-decision binary digit together with 2 binary digits of information about the quality of this hard decision. Almost all coding systems that have been made for the deep-space channel and similar channels have operated with such 3-bit soft decisions.

2.3 Bandwidth Considerations

We have already observed that the user of the deep-space channel has virtually unlimited bandwidth at his disposal. This does not mean, of course, that he should use as much bandwidth as possible. Rather, analogous to Fano's dictum for demodulator quantization, he should in fact use as little bandwidth as possible consistent with not unduly decreasing the capacity, C_∞, of the original waveform channel. One reason for this is that the receiver's radio-frequency "front-end" must have as wide a bandwidth as the transmitted signal and, when this bandwidth becomes too great, the large Gaussian noise present at the receiver's front-end makes it virtually impossible to realize the coherent demodulation that is required to reap the theoretically available benefits of greater bandwith--a kind of "Catch-22" of large bandwidth. We will presently see another cogent and related reason for using as little bandwidth as possible consistent with not unduly decreasing capacity.

The *code rate* R (bits/digit) of a binary channel code is the average number of information bits per binary digit produced by the code (on a long-time basis). The minimal requirement that the encoding be invertible specifies that $R \leq 1$, where $R = 1$ corresponds to uncoded transmission. Suppose the code is used with binary modulation so that each encoded binary digit selects one modulation symbol. Then, because r is the number of modulation symbols per second, $\mathcal{R} = r \times R$ is the information transmission rate in bits per second and is the number that is specified in advance to the designers of the communication system. But the Fourier bandwidth of the resulting sequence of waveforms is directly proportional to $r = \mathcal{R}/R$ rather than to $\mathcal{R}$ itself, which leads to the inescapable conclusion that when designing a coding system for the deep-space channel, one should choose the minimum code rate consistent with not unduly decreasing the capacity per second of the corresponding discrete channel from its value C_∞ for the infinite-bandwidth waveform channel.

2.4 Fourier Bandwidth and Shannon Bandwidth

To proceed further with our consideration of bandwidth for the deep-space channel, it is helpful to make the important distinction between Fourier bandwidth and Shannon bandwidth. Shannon [1], [2] in fact identified bandwidth with the number of signal-set dimensions that are transmitted per second; we shall use the symbol B to denote this quantity that we will call the *Shannon bandwith* of the transmitted signal. If the modulation signal set is n-dimensional, then $B = r \times n$, as follows from the fact that r is the number of modulation signals transmitted per second. Shannon's equation (1) can be written more fundamentally in terms of Shannon bandwidth as

$$C_B = \frac{1}{2} B \log_2 (1 + \frac{S}{B N_0/2}) \quad \text{(bits/sec)}. \tag{9}$$

The consistency between (1) and (9) can be seen as follows. According to the Shannon-Nyquist sampling theorem (i.e., Shannon's completion [1] of the theorem begun by Nyquist [7]), at most 2W orthogonal signals can be sent per second in a frequency band of Fourier bandwidth W. Thus, $B \leq 2W$ with equality if the orthogonal signals are translates by $1/(2W)$ seconds of the familiar sinc signal that has a flat spectrum over the given frequency band, i. e., if the transmitted signals fill the available Fourier bandwidth as completely as possible. Using this maximum $B = 2W$ in (9) yields (1), as indeed it must because C_B as given by (9) increases monotonically with B and hence, if a constraint W on the Fourier bandwidth is given, one must choose the maximum Shannon bandwidth B consistent with that constraint. Examination of Shannon's arguments in [1] show in fact that he first proved the capacity formula (9), then made use of the sampling theorem to deduce (1). Equation (9) is indeed the more fundamental of these two capacity formulas since it holds regardless of whether or not one chooses modulation signals that completely fill out the Fourier bandwidth. Communications engineers must live with the constraints on Fourier bandwidth specified by regulatory agencies, but the performance of their systems depends much more on their Shannon bandwidth rather than on their Fourier bandwidth.

We are finally in position to see how the code rate R affects the capacity of the deep-space channel. Because binary antipodal signalling uses a one-dimensional (n = 1) signal set, we have $B = r$ and thus Shannon's capacity formula (9) becomes

$$C_r = \frac{1}{2} r \log_2 (1 + \frac{S}{r N_0/2}) \quad \text{(bits/sec)}. \tag{10}$$

Recalling that $S = r \times E = C_\infty \times R \times E_b$, we see that (10) can be rewritten as

$$C_r = \frac{1}{2} \frac{S}{R E_b} \log_2 (1 + \frac{R E_b}{N_0/2}) \quad \text{(bits/sec)}. \tag{10}$$

Letting the code rate R tend to 0, we see that C_r tends to the limit C_∞ given by (2)--as of course it must since, for fixed signal power S and fixed energy per information bit E_b, $B = r$ tends to infinity as R tends to 0. Moreover, we see that, for any given non-zero rate R, the capacity reduction factor $\gamma = C_r/C_\infty$ due to the resulting finite bandwidth is given by

$$\gamma = \frac{ln\left(1 + \dfrac{R\,E_b}{N_0/2}\right)}{\dfrac{R\,E_b}{N_0/2}} \, . \tag{11}$$

Suppose next that we are operating near the Shannon limit, i. e., that $E_b/N_0 \approx ln\,2 \approx 0.69$. The capacity reduction factor γ then becomes

$$\gamma \approx \frac{ln\,(1 + 1.386\,R)}{1.386\,R} \, , \tag{12}$$

which is our desired result for specifying how much loss will be suffered due to finite bandwidth by a coding system that operates near the Shannon limit.

For a code rate $R = 1/2$, which is the rate that was selected for the Pioneer 9 channel coding system, equation (12) shows a capacity reduction factor $\gamma = 0.76$ (-1.2 dB) as the penalty paid for the resulting finite bandwidth. For a code rate $R = 6/32$, which is the rate of the Mariner '69 channel code, the reduction factor is only $\gamma = .889$ (-0.51 dB). One might suppose then that the Mariner '69 code rate was a wiser choice than the Pioneer 9 code rate. In fact, however, the Mariner '69 code was chosen for reasons, which will be explained below, that had nothing to do with minimizing the capacity loss due to finite bandwidth. It would, in fact, have been more desirable to use the code rate $R = 1/2$ in spite of the 0.7 dB increased capacity loss, since the symbol energy $E = R \times E_b$ at rate $R = 6/32$ is so much smaller than at $R = 1/2$ (4.2 dB smaller for the same E_b) that the phase-lock loops that are used to perform coherent demodulation of BPSK signals tend to lose lock unacceptably often when the lower code rate is used. One of the lessons of these first applications of channel coding in deep-space was that the use of an energy-efficient channel coding system places unusually severe demands on the phase-tracking loops in the demodulator because of the very low energy E of the transmitted waveforms.

3 A Tale of Two Codes

We will now take a closer look at the specific channel codes that were chosen for the Pioneer 9 and Mariner '69 downlink channels. Our discussion in the previous section has already told us two important considerations for these channel coding systems, viz.,

• The decoder must make use of soft-decisions on the transmitted digits.

• The code rate should be at most 1/2, but preferably as near 1/2 as possible.

3.1 The Mariner '69 Code

The first choice that had to be made by the designers of the Mariner '69 channel coding system was whether to use a block code or a convolutional code. Several factors militated against the choice of a convolutional code. At the time that the decision on the Mariner '69 code had to be made, the only general soft-decision decoding technique for convolutional codes that was known was Wozencraft's original sequential decoding algorithm, cf. [8]. The much more efficient and easily implemented Fano algorithm [9] for sequential decoding was still in the process of discovery and development. Wozencraft's algorithm had already been implemented in special purpose hardware at the M.I.T. Lincoln Laboratory, but for application on telephone channels (for which it turned out to be not well suited) rather than for the deep-space channel (for which it would have been well suited). Memory in a channel tends to cause large increases in the computation required to do sequential decoding; the deep-space channel is memoryless but the telephone channel is certainly not. In any case, there was no convincing evidence at the time that when the decision on the Mariner '69 code had to be made that sequential decoding would be a good and practical choice. The Mariner '69 code designers at the CalTech Jet Propulsion Laboratory (JPL) in Pasadena, California, opted for a block code, a decision that we find difficult to fault even in retrospect.

Unfortunately, there was at that time only a few block codes for which a practical soft-decision decoding algorithm was known--a situation that has not changed appreciably in the intervening 30 years! Moreover, these codes generally had very low code rates, which as we have seen is fine for energy-efficiency on the deep-space channel but nonetheless undesirable because of the heavy demand that the expanded bandwidth places on the phase-lock loops required for coherent demodulation of BPSK signals. The most attractive block codes available for practical soft-decision decoding were the first-order Reed-Muller codes [6]. These first-order Reed-Muller codes are binary parity-check codes with blocklength $n = 2^m$, minimum Hamming distance $d_{min} = 2^{m-1}$ and $k = m + 1$ information bits, where $m \geq 2$ is a design parameter. The code rate, $R = k/n = (m + 1)/(2^m - 1)$, is unpleasantly small except for small m. These codes can be viewed in many different ways but, when used with binary antipodal signalling, they are perhaps best seen as realizing the "biorthogonal signal set" in n dimensions, cf. [3, p. 261].

The n-dimensional "orthogonal signal set of energy £" is a set of n vectors $x_1, x_2, ... , x_n$ in n-dimensional Euclidean space with the property that each pair of vectors is orthogonal (i.e., the inner product between each pair of vectors vanishes) and each vector has squared norm £ (i. e., its innner product with itself is £). The simplest construction of this signal set is to choose the n vectors with a single non-zero component equal to $\sqrt{E}$, from which it is easy to see that the squared Euclidean distance $(d_E)^2$ between any two signals is $2 \times E$, but many other constructions are possible. For instance, with n = 4, the rows of the Hadamard matrix

$$H_2 = \begin{bmatrix} +1 & +1 & +1 & +1 \\ +1 & -1 & +1 & -1 \\ +1 & +1 & -1 & -1 \\ +1 & -1 & -1 & +1 \end{bmatrix},$$

$$(13)$$

when scaled by $\sqrt{E/4}$, yield the 4-dimensional orthogonal signal set of energy E. The *n-dimensional biorthogonal signal set of energy* E is the set of $2 \times n$ signals formed by augmenting the orthogonal signal set with the negative of each of its signals. Each signal is at squared Euclidean distance $(d_E)^2 = 2 \times E$ from all the other signals excepts its negative, from which it is at squared Euclidean distance $(d_E)^2 = 4 \times E$.

Consider now sending a signal from any set of equi-energy signals in n-dimensional Euclidean space over a channel such that the received vector r is the sum of the transmitted signal and an additive noise vector n whose components are independent Gaussian random variables, all with zero mean and the same variance. The maximum-likelihood detection rule (which is the rule that minimizes the error probability in choosing the transmitted signalwhen all signals are equally likely) is to choose that signal x_i whose inner product (or "correlation") with r is greatest, cf. [3, p. 234]. For the "Hadamard signal set" of the preceding paragraph, we see that $H_2 r$ is (within an unimportant positive scale factor) the vector of correlations between r and each of the signals in the signal set. Thus, the maximum-likelihood detection rule reduces to: Choose the signal x_i corresponding to the location i of the maximum component of $H_2 r$. Decoding the corresponding biorthogonal signal set is almost as simple--the maximum-likelihood detection rule becomes: Choose the signal to be x_i or $-x_i$, where i is the location of the maximum-magnitude component of $H_2 r$, according as to whether this component is positive or negative, respectively.

The relationship of the above to the Mariner '69 channel code stems from the fact that the first-order Reed-Muller code of length n = 2^m, when the binary codewords are mapped into vectors in n-dimensional Euclidean space in the manner that a binary 0 is mapped into $+\sqrt{E}$ and a binary 1 is mapped into $-\sqrt{E}$, yields precisely the biorthogonal signal set corresponding to the $2^m \times 2^m$ Hadamard matrix H_m, where the Hadamard matrices are defined recursively by

$$H_m = \begin{bmatrix} H_{m-1} & H_{m-1} \\ H_{m-1} & H_{m-1} \end{bmatrix},$$

$$(14)$$

in the same manner as described in our example with H_2. It follows that this code can be decoded with maximum-likelihood soft-decision decoding on the deep-space channel by first

13

mapping the received waveform over the corresponding n modulation symbol intervals to the vector r of coefficients in the representation (of the "relevant" portion) of this waveform as a linear combination of the n orthonormal waveforms obtained by translation of the basis waveform for the one-dimensional modulation signal set, then using the optimum detection rule described above, which reduces essentially to the computation of $H_m r$. Since the entries of the matrix H_m are all either +1 or -1, the obvious calculation of $H_m r$. would take a total of $n(n-1) \approx n^2$ additions and subtractions. But this matrix multiplication is precisely the rule of calculation for the so-called *Walsh-Hadamard transform* of r, for which there exists a fast-transform method that requires only $n \, log_2 \, n$ additions and subtractions. All this was well known in the early 1960's to the engineers at JPL, who built a special purpose digital device that they called the "Green machine" (not after its color but after its designer) to perform the fast Walsh-Hadamard transform on vectors r of length $n = 2^5$. This device then served as the decoder for the (n = 32, k = 6) Reed-Muller code used in the Mariner '69 spacecraft, cf. [11].

It remains to see why this specific Reed-Muller code was the one selected. The probability of error in deciding between two equi-energy vectors in additive Gaussian noise (whose components are independent Gaussian random variables with zero means and variances $N_0/2$) depends only on the squared Euclidean distance between these signals and is given by $Q(\sqrt{(d_E)^2/2N_0}$, [as can be inferred from our discussion preceding equation (7) above where $(d_E)^2 = 4E_b$]. If a binary code has minimum Hamming distance d_{min} and its binary codewords are mapped into vectors in n-dimensional Euclidean space in the manner that a binary 0 is mapped into $+\sqrt{E}$ and a binary 1 is mapped into $-\sqrt{E}$, then the minimum squared Euclidean distance between the resulting vectors is

$$(d_{Emin})^2 = 4 E d_{min.} \tag{15}$$

Thus, the probability of a decoding error will be given by $P_e \approx Q(\sqrt{2 E d_{min}/N_0})$, where we have neglected to account for the multiplicity of vectors that may be at this same minimum squared Euclidean distance from the transmitted codeword. But $E = RE_b$, where R is the code rate, so

$$P_e \approx Q(\sqrt{2 R E_b d_{min}/N_0}). \tag{16}$$

Comparing (7) and (16) shows that, at least to the first order, the coding gain G_c is given by

$$G_c \approx R d_{min,} \tag{17}$$

which is a very useful formula for evaluating binary codes used with soft-decision decoding and binary antipodal signalling on the deep-space channel. Using the parameters of the first-order Reed-Muller codes in (17) gives

$$G_c \approx \frac{m+1}{2}. \tag{18}$$

This shows that it is desirable to choose m as large as possible, i. e., up to the point where the phase-lock loop tracking problem induced by the bandwidth increase can still be overcome. The JPL engineers chose m = 5, for which (18) gives an estimated gain of $G_c \approx 3$ (4.8 dB). The actual gain is somewhat smaller because of the large multiplicity of nearest neighbors in the first-order Reed-Muller codes--there are 62 nearest neighbors for each codeword in the m = 5 code. The actual gain was 2.2 dB at a bit error probability of 5×10^{-4} [Posner gives the figure as 5×10^{-3} in [11] but this is apparently incorrect; the figures given in [11] for the uncoded Mariner IV system (an E/N_0 of 8.5 dB), which had the same bit error probability as the Mariner '9 system, yield a bit error probability of 5×10^{-4}, as can be checked from equation (7).]

3.2 The Pioneer 9 Code

The designers of the Pioneer 9 channel coding system had the advantage of starting work in the mid-1960's when the power and practicality of the Fano sequential decoding algorithm [9] were becoming well known. With little hesitation, these designers settled on a rate R = 1/2 convolutional coding system to be decoded by the Fano algorithm. Sequential decoding in general is a technique by which the decoder works (at least in principle) with the tree of partial encoded sequences that it has previously examined for their "likelihood" with respect to the received sequence, extending each time the most recent explored sequence until its "likelihood" falls below some threshold. Fano in 1964 [9] had made two important contributions to sequential decoding. First, he introduced a "metric" that on intuitive grounds should correspond to a reasonable notion of "likelihood" when applied to possible partial encoded sequences of different length. [It was not until many years later that we were able to prove [12] that this "Fano metric" was precisely the metric that yields the true maximum-likelihood decision as to which partially explored encoded sequence to extend. This delayed theory is illustrative of the theoretical intricacy of sequential decoding!] Second, he introduced his own extension algorithm, which is the soul of simplicity and, at the same time, the most subtle algorithm that this writer has ever seen. Fano's algorithm uses almost no memory, which was an important consideration in the mid-1960's, and remains today the fastest sequential decoding algorithm known.

The main problem with sequential decoding is the variability of the decoding computation. Generally (but not always) when sequential decoding is used, the output of the convolutional encoder is segmented into independent "frames" by occasionally inserting a pattern of M consecutive 0's into the information bit stream to drive the encoder back to the all-zero initial state. This yeilds a finite (but still very large) potential code tree to be explored for each frame. How long it takes the decoder to work its way to the end of this code tree depends on how "noisy" the actual received frame is.

The code that was used in the Pioneer 9 system was a rate $R = 1/2$ non-systematic convolutional code that had been constructed by S. Lin and H. Lyne, cf. [13, p. 539]. [The term "non-systematic" refers to the fact that the information bits do not appear unchanged among the encoded digits--it is really the encoder that should be called "non-systematic" rather than the code. In this paper we have followed the usual communications practice of not distinguishing between a "code" (i. e., the set of all codewords) and an "encoder," but this difference can be important in coding theory.] The Lin-Lyne code had a memory of $M = 20$, i. e., the encoder remembered twenty past information bits as well as the current information bit when forming the two encoded binary digits that are emitted for each input information bit. It was known that the minimum Hamming distance d_{min} of this code was 11. This is the minimum distance between two encoded sequences of length $2 \times (M+1) = 22$ digits (one "constraint length") that correspond to different values of the initial information bit. If one uses these code parameters mindlessly in (17), one computes an estimated coding gain of $G_c \approx$ 5 (7.0 dB), but this estimate is so optimistic that it must be taken with a grain of salt! [In fact, one should really use in (17) the *free distance*, d_{free}, of the convolutional code, which is the minimum Hamming distance between two encoded sequences of infinite length that correspond to different values of the initial information bit. This would, of course, give an even more optimistic estimate of G_c!] The problem is not one of great multiplicity of near neighbors as it was for the Reed-Muller code. Rather, the problem is that one must take into account the fact that although Fano algorithm sequential decoding is virtually maximum-likelihood decoding *if the decoder is allowed to compute until it makes a decoding decision*, in practice *one always aborts the decoding after some predetermined amount of computation on a frame and announces erasure of the corresponding frame of data*. The actual error probability in the non-erased frames is virtually zero since the frames that would have resulted in decoding errors with maximum-likelihood decoding are generally frames that would require enormous computation to decode. [This latter feature was attractive to the scientists with experiments aboard Pioneer 9--they could really trust any experimental data radioed back from Pioneer 9 that was not erased by the sequential decoder.] The way that computation increases with decreasing signal-to-noise ratio is the primary determiner of the actual signal-to-noise ratio at which one can operate with sequential decoding, and hence of the actual coding gain. For the Pioneer 9 system, the actual gain was about 3.0 dB, cf. [13, p. 539].

3.3 Comparison of the Two Codes and Why the Pioneer 9 Code Was the First into Space

On all counts, the Pioneer 9 channel coding system was better than the Mariner '69 channel coding system. It offered greater coding gain (3.0 dB vs. 2.2 dB) and at the same time its higher rate (R = 1/2 vs R = 6/32) meant considerably less bandwidth expansion and hence much better tracking by the phase-lock loops in the coherent demodulator. The Fano-algorithm sequential decoder for the Pioneer 9 system was essentially cost-free--it was realized in software during the spare computation time of an already on-site computer, whereas the "Green machine" for decoding the Mariner '69 code was a non-negligible piece of electronic hardware.

The fact that the superior Pioneer 9 channel coding system got into space sooner than the inferior Mariner '69 channel coding system, even though the former was designed several years later than the latter (which is why it was a better system), is due primarily to the efforts of a single person, D. R. Lumb of the NASA Ames Research Center. Lumb had been quick to appreciate the importance of Fano-algorithm sequential decoding for the deep-space channel. In this, he was influenced by G. D. Forney, Jr., of Codex Corporation and by two Codex consultants, R. G. Gallager and (to a much lesser extent) this writer. After rapid development of the convolutional coding system, Lumb succeeded in getting it aboard Pioneer 9 as *an experiment*. This neatly side-stepped the long approval time that would have been necessary if this coding system had been specified as part of an operational communications system for a spacecraft. The operational communications system for Pioneer 9 was, of course, an uncoded BPSK system. The experimental coding system was activated as soon as Pioneer 9 was launched--it was never turned off!

References

[1] C. E. Shannon, "A Mathematical Theory of Communication," *Bell Sys. Tech. J.*, vol. 27, pp. 379-423 and 623-656, July and Oct. 1948.

[2] C. E. Shannon, "Communication in the Presence of Noise," *Proc. IRE*, vol. 37, pp. 10-21, Jan. 1949.

[3] J. M. Wozencraft and I. M. Jacobs, *Principles of Communication Engineering.* New York: Wiley, 1965.

[4] R. M. Fano, *Transmission of Information.* Cambridge, Mass.: MIT Press and Wiley, 1961.

[5] J. L. Massey, "Coding and Modulation in Digital Communications," pp. E2(1)-E2(4) in *Proc. Int. Zurich Seminar*, Zürich, Switzerland, March 1974.

[6] I. M. Jacobs, "Sequential Decoding for Efficient Communication from Deep Space," *IEEE Trans. Commun. Tech.*, vol. COM-15, pp. 492-501, August 1967.

[7] H. Nyquist, "Certain Factors Affecting Telegraph Speed," *Bell Sys. Tech. J.*, vol. 3, pp. 324-346, April 1924.

[8] J. M. Wozencraft and B. Reiffen, *Sequential Decoding.* Cambridge, Mass.: MIT Press and Wiley, 1961.

[9] R. M. Fano, "A Heuristic Discussion of Probabilistic Decoding," *IEEE Trans. Info. Th.*, vol. IT-9, pp. 64-73, April 1963.

[10] I. M. Reed, "A Class of Multiple-Error-Correcting Codes and the Decoding Scheme," *IRE Trans. Info. Th.*, pp . 38-49, Sept. 1954.

[11] E. C. Posner, "Combinatorial Structures in Planetary Reconnaissance," pp. 15-46 in *Error Correcting Codes* (Ed. H. B. Mann). New York: Wiley, 1968.

[12] J. L. Massey, "Variable-Length Codes and the Fano Metric," *IEEE Trans. Info. Th.*, vol. IT-18, pp. 196-198, Jan. 1972.

[13] S. Lin and D. J. Costello, Jr., *Error Control Coding: Fundamentals and Applications.* Englewood Cliffs, NJ: Prentice-Hall, 1983.

Coding and Coded Modulation Techniques
for
Reliable Satellite and Space Communications [1]

Shu Lin, Sandeep Rajpal and Do Jun Rhee

Department of Electrical Engineering

University of Hawaii at Manoa

Honolulu , Hawaii 96822, U.S.A.

ABSTRACT

This paper presents two coding and coded modulation schemes for reliable data transmission with high spectral efficiency, large coding gain, and reduced decoding complexity. In the first scheme, coded modulation is used in conjunction with concatenation either in single level or in multiple levels. In the second scheme, multilevel coded modulation is combined with multiple product codes. Three specific concatenated coded modulation schemes and one product coded modulation scheme are presented. Analytical and simulation results show that these schemes provide high reliability and achieve large coding gains over the uncoded QPSK system with reduced decoding complexity. These schemes have been proposed for the new NASA Bandwidth Efficient Standard for high-speed satellite communications.

[1] This research was supported by NASA Grant NAG 5-931

I. Introduction

This paper presents two coding and coded modulation schemes for achieving reliable data transmission with large coding gain, high spectral efficiency, and reduced decoding complexity. In the first scheme, coded modulation[1] is used in conjunction with concatenation[2]. This combination of coded modulation and concatenation is known as concatenated coded modulation[3]. In concatenated coded modulation schemes, the concatenation can be carried out either in single level or in multiple levels. The inner codes are bandwidth efficient modulation codes(block or trellis), and the outer codes are generally block codes with code symbols from some Galois fields. If the inner codes, outer codes, and the level of concatenation are properly chosen, extremely good error performance can be achieved with reduced decoding complexity, high spectral efficiency, and large coding gain. In the second scheme, multilevel coded modulation[4-7] is combined with product codes. If the component codes of the product codes are properly chosen, long powerful modulation codes of high spectral efficiency can be constructed. These codes have multilevel structure and can be decoded with suboptimum multi-stage decoding which achieves good error performance with reduced decoding complexity.

In this paper, several specific concatenated and product coded modulation schemes are presented. These schemes have been proposed for the new NASA Bandwidth Efficient Coding Standard for high-speed satellite communications. In the proposed concatenated coded modulation schemes, the inner codes are 8-PSK modulation codes and the outer codes are Reed-Solomon(RS) codes. The inner codes are chosen to have simple trellis structure and they are decoded with the soft-decision Viterbi decoding algorithm either in single stage or in multiple stages. In the proposed product coded modulation schemes, multilevel block modulation codes are used as the vertical codes, binary BCH codes are used as the horizontal codes, and decoding is carried out in multiple stages.

2. Single-Level Concatenated Coded Modulation Schemes

2.A Scheme-I

The schematic diagram of this scheme is shown in Figure 1.

Outer Code

The outer code B is the NASA standard (255, 223) RS code with code symbols from $GF(2^8)$. Each code symbol is represented by an 8-bit byte. The code is capable of correcting up to 16 symbol errors. In the scheme, the code is interleaved to a depth of $m=2$.

Inner code

The inner code A is a block 8-PSK modulation code of length $n=8$. The code is formed by combining three binary codes, denoted A_{b1}, A_{b2} and A_{b3}, through proper bits-to-signal mapping where: (1) A_{b1} is the (8, 1) **repetition** code with minimum Hamming distance 8; (2) A_{b2} is the (8, 7) **even-parity check** code with minimum Hamming distance 2; and (3) A_{b3} is the (8, 8) **universal** code with minimum Hamming distance 1. To construct the modulation code A, we first label each signal point of the 8-PSK signal constellation by three bits, abc, as shown in Figure 2, where a is the first labeling bit and c is the last labeling bit. We see that: (1) Two signal points with labels different at the first bit position are at a **squared Euclidean distance** at least $d_1=0.586$ apart; (2) Two signal points with labels identical at the first bit position but different at the second bit position are separated by a squared Euclidean distance at least $d_2=2$ apart; and (3) Two signal points with labels identical at the first two bit positions but different at the last bit position are at a squared Euclidean distance $d_3=4$ apart. Let

$$\mathbf{a} = (a_1, a_2, a_3, a_4, a_5, a_6, a_7, a_8)$$

$$\mathbf{b} = (b_1, \ b_2, b_3, b_4, \ b_5, b_6, b_7, \ b_8)$$

$$\mathbf{c} = (c_1, \ c_2, c_3, c_4, \ c_5, c_6, c_7, \ c_8)$$

be three codewords in A_{b1}, A_{b2} and A_{b3} respectively. We form the following sequence:

$$\mathbf{a} * \mathbf{b} * \mathbf{c} \triangleq (a_1 b_1 c_1, \ a_2 b_2 c_2, \ \cdots, \ a_8 b_8 c_8)$$

For $1 \leq i \leq 8$, we take $a_i b_i c_i$ as the label for a signal point in the 8-PSK signal constellation as shown in Figure 2. Let $\lambda(\cdot)$ be the mapping which maps the label $a_i b_i c_i$ into its corresponding signal point s_i, i.e., $\lambda(a_i b_i c_i) = s_i$. Then

$$\lambda(\mathbf{a} * \mathbf{b} * \mathbf{c}) = (\lambda(a_1 b_1 c_1), \ \lambda(a_2 b_2 c_2), \ \cdots, \ \lambda(a_8 b_8 c_8))$$

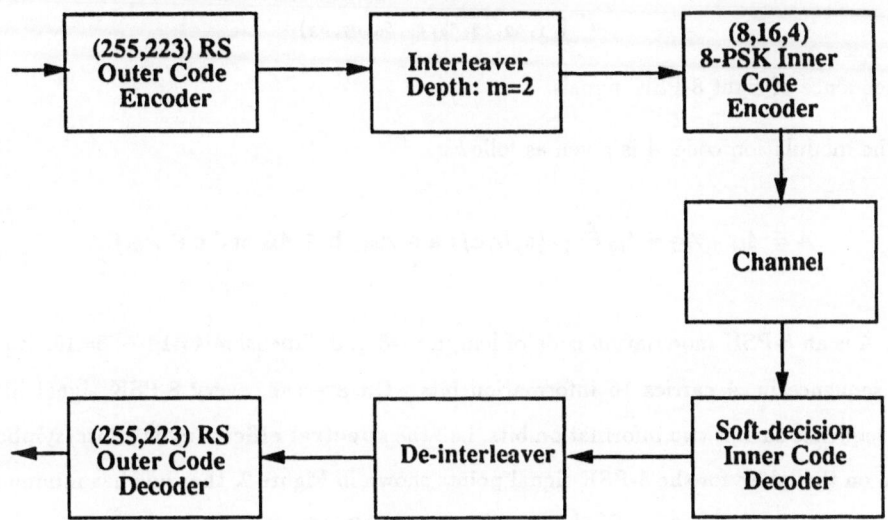

Figure 1   A single-level concatenated coded modulation system with the
            NASA standard (255,223) RS code as the outer code and a
            (8,16,4) 8-PSK code as the inner code

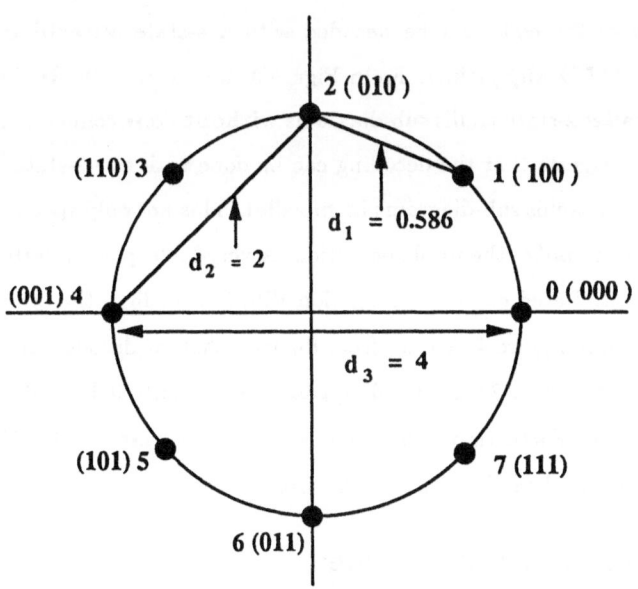

Figure 2   An 8-PSK signal constellation and its signal labels

$$= (s_1, s_2, s_3, s_4, s_5, s_6, s_7, s_8)$$

is a sequence of eight 8-PSK signals.

The modulation code A is given as follows:

$$A \triangleq A_{b1} * A_{b2} * A_{b3} \triangleq \{\lambda(\mathbf{a}, \mathbf{b}, \mathbf{c}) : \mathbf{a} \in A_{b1}, \ \mathbf{b} \in A_{b2} \text{ and } \mathbf{c} \in A_{b3}\}.$$

Then A is an 8-PSK modulation code of length $n=8$ and dimension $k=1+7+8=16$. Each code sequence in A carries 16 information bits. On average, every 8-PSK signal in a code sequence carries two information bits, i.e., the **spectral efficiency** is 2 bits/symbol. Based on the labels for the 8-PSK signal points shown in Figure 2, the code has minimum squared Euclidean distance, $D[A] = \min\{8 \times 0.586, \ 2 \times 2, \ 1 \times 4\} = 4.$[5 − 7] For convenience, we denote this code with (8, 16, 4), i.e., length 8, dimension 16, and minimum squared Euclidean distance 4. This code achieves a 3 dB **asymptotic coding** gain over the **uncoded** QPSK with the same spectral efficiency, 2 bits/symbol.

The above inner code A has a 4-**state** 8-**section trellis diagram** as shown in Figure 3[3]. Hence, the code can be decoded with a 4-state **Viterbi decoder** using a **soft-decision MLD algorithm**. From Figure 3, we see that the trellis consists of two **identical parallel** 2-state trellis sub-diagrams **without** cross connections between them. This structure suggests that the decoding can be done with two 2-state Viterbi decoders to process the two trellis sub-diagrams **in parallel**. This not only speeds up the decoding process but also simplifies the implementation. A very high speed Viterbi decoder for this code can be built. Based on the soft-decision Viterbi decoding, the error performance of the code is shown in Figure 4. We see from Figure 4 that, at decoded bit-error-rate(BER) 10^{-6}, this code achieves a 2.0 **real** coding gain over the uncoded QPSK **without bandwidth expansion**. Furthermore, the code is **phase invariant** under 45° rotation. This property ensures rapid carrier-phase resynchronization.

Overall Concatenated Scheme

The encoding of the proposed scheme is performed in two stages as shown in Figure 1. First a message of 223×8 bits is divided into 223 8-bit bytes. Each 8-bit byte is regarded

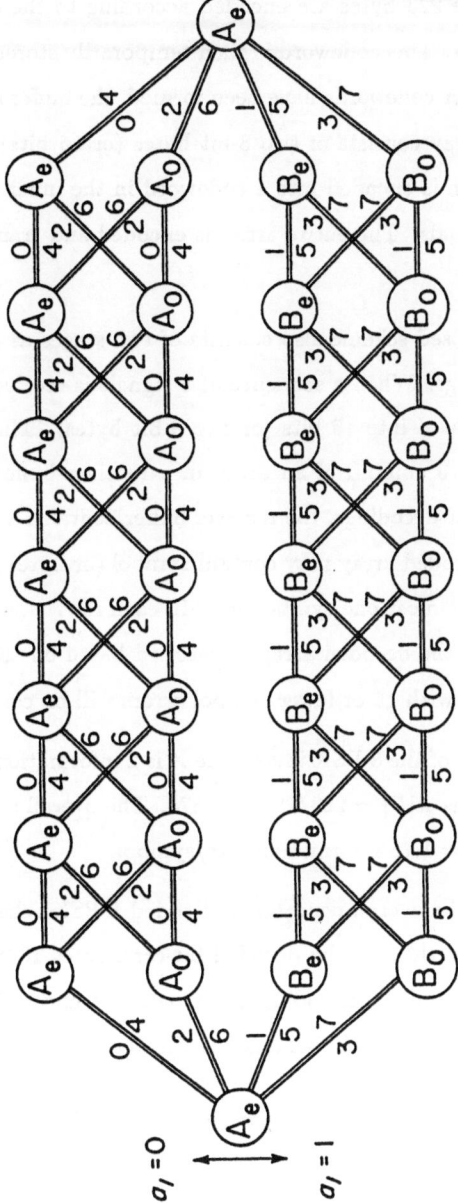

Figure 3 A 4-state trellis diagram for the (8, 16, 4) 8-PSK code used in the first concatenated coded modulation scheme

as a symbol in $GF(2^8)$. These 223 bytes are encoded according to the outer code B to form a 255-byte codeword in B. This codeword is then temporarily stored in a buffer as a row in an array. After two outer codewords have been formed, the buffer stores an 2×255 array. Each column of the array consists of two 8-bit bytes (or 16 bits). At the second stage of encoding, each column is encoded into a codeword in the inner code A which is a sequence of eight 8-PSK signals. The entire array is encoded and transmitted column by column.

The decoding for the proposed scheme also consists of two stages as shown in Figure 1, the inner and outer decodings. When a sequence of 8 signals is received, it is decoded based on the 8-PSK inner code A into 16 bits (or two 8-bit bytes). These decoded two 8-bit bytes are then stored as a column of an array in a receiver buffer for the second stage decoding. After 255 inner decodings, the receiver buffer contains a 2×255 decoded array. Each column of this decoded array may contain symbol (or byte) errors which are distributed among two rows, at most one symbol error in each row. Now second stage of decoding begins. Each row of the decoded array is decoded based on the (255, 223) RS outer code. Any error pattern with 16 or fewer symbol errors will be corrected.

Since the spectral efficiency of the 8-PSK inner code A is 2 information bits/signal and the rate of the outer RS code is $R[B] = 223/255 = 0.875$. The overall spectral efficiency of the system is $\eta = 2 \times R[B] = 1.75$ information bits/symbol.

The error performance of the overall concatenated coded 8-PSK scheme is simulated and shown in Figure 5. We see that, at the decoded bit-error-rates 10^{-6} and 10^{-10}, the scheme achieves 5.12dB and 7.10dB real coding gains over the uncoded QPSK system respectively. For SNR $E_b/N_o = 6.2$ dB the scheme practically provides **error-free data transmission**.

2.B Scheme-II

In the second proposed concatenated coded modulation scheme, the outer code B is again the NASA standard (255 , 223) RS code. However, the inner code is a 4-state trellis 8-PSK code.

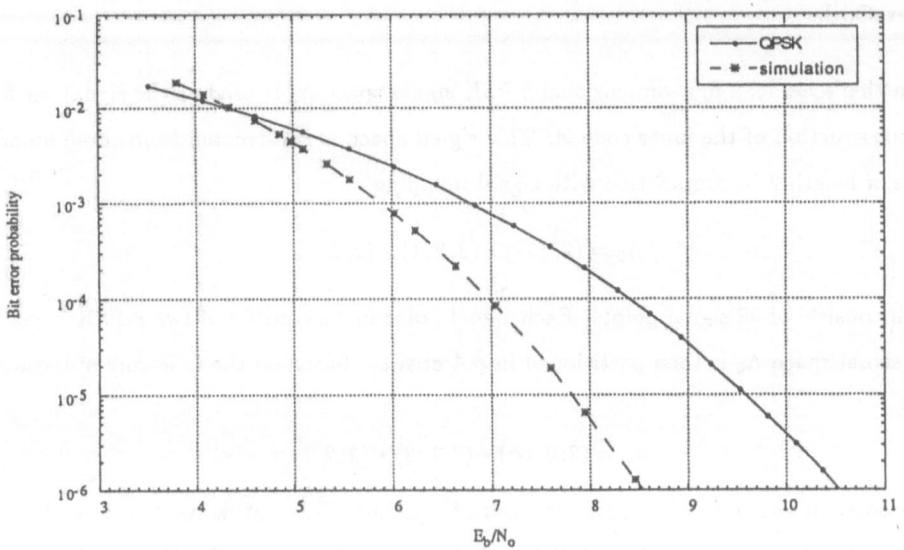

Figure 4   Bit error performance of the (8, 16, 4) 8-PSK code

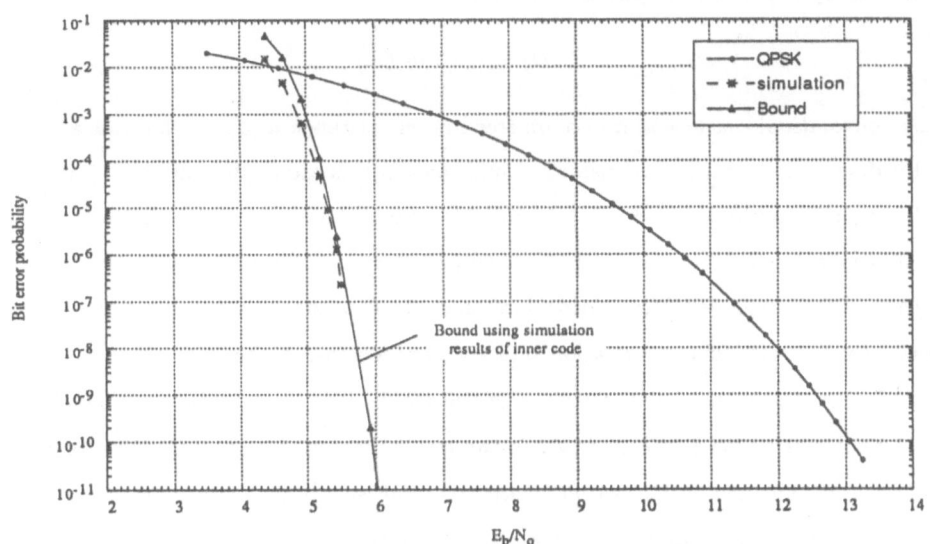

Figure 5   Bit error performance of the first concatenated coded
8-PSK scheme

Inner Code

In this scheme, a four-dimensional 8-PSK signal space Λ_0 is used as the signal set for the construction of the inner code A. This signal space is constructed from three binary codes of length 2 in conjunction with signal mapping :

$$\Lambda_0 = (2,1,2) * (2,2,1) * (2,2,1)$$

which consists of 32 signal points. Each signal point in Λ_0 consists of two 8-PSK signals. The signal space Λ_0 is then partitioned into 4 **cosets** based on the following **subspace** of Λ_0 :

$$\Lambda_1 = (2,0,\infty) * (2,1,2) * (2,2,1)$$

Each coset in the partition Λ_0/Λ_1 consists of 8 points. The **intra-set (squared Euclidean)** distance of Λ_0 is $\Delta_0 = 1.172$, and the intra-set distance of each coset in Λ_0/Λ_1 is $\Delta_1 = 4.0$.

The trellis 8-PSK inner code A is formed from a specific rate - 1/2 convolutional code and the signal space Λ_0 as shown in Figure 6. The rate- 1/2 convolutional code is generated by the following generator matrix :

$$G(D) = \left(D, 1 + D^2 \right)$$

This convolutional code has minimum free-branch distance $d_{\text{B.free}} = 3[8]$ and a 4-state trellis diagram. At each time instant, 4 information bits are applied at the input of the trellis coded modulation encoder (Figure 6). One information bit is encoded by the convolutional encoder into two coded bits which select one of the 4 cosets in the partition Λ_0/Λ_1. The 3 uncoded information bits then select a signal point from the selected coset. As a result, four information bits are encoded into two 8-PSK signals. All the possible signal sequences at the output of the trellis coded modulation encoder form a **4-dimensional** trellis 8-PSK code A. In the trellis diagram for the code, two **adjacent states** are connected by 8 **parallel branches** and each branch corresponds to a signal point (or two 8-PSK signals) in a coset in the partition Λ_0/Λ_1. The spectral efficiency of the code is $\eta = 2$ bits/signal. The code is **invariant** under 90° phase rotation [8]. The 4-dimensional trellis 8-PSK code A has **free** squared Euclidean distance equal to 4 and achieves 3 dB **asymptotic coding gain** over the uncoded QPSK system. With the soft-decision Viterbi **decoding algorithm**, the error performance of this code is shown in Figure 7. From Figure 7, we see that, at the decoded BER 10^{-6}, the code achieves a 2.65 dB **real coding gain** over the encoded QPSK system **without** bandwidth expansion.

Overall Concatenated Scheme

The overall concatenated scheme is shown in Figure 8. The outer RS code is interleaved to a depth of $m = 4$ as shown in Figure 9. Four RS codewords form an **array** with 255 columns, each column consists of four 8-bit bytes. The array is encoded based on the trellis 8-PSK inner code and transmitted **column by column**. Each 8-bit byte in a column is divided into two 4-bit **blocks**. Each 4-bit block is encoded into two 8-PSK signals based on the trellis 8-PSK inner code. The **order** of encoding of the eight 4-bit blocks in a column follows the integer sequence shown in Figure 9. The **overall spectral efficiency** of the scheme is $\eta = 1.75$ bits/symbol (same as the first scheme).

As usual, the decoding of the overall scheme consists of two stages, the inner and outer decodings. The trellis 8-PSK inner code is decoded first with the soft-decision Viterbi decoding algorithm. A decoded sequence of $255 \times 8 = 2040$ 4-bit blocks at the output of the inner code decoder is de-interleaved and rearranged into an array of 4 rows and 255 columns. Each row is then decoded based on the RS outer code.

To reduce the encoding and decoding delay, 4 **pairs** of inner code encoder/decoder can be used. Since the trellis inner code is chosen to have **low decoding complexity**, the duplication adds just **a little more cost**.

The overall concatenated coded 8-PSK scheme has been analyzed and an upper bound on the bit error probability has been derived. The error performance of the scheme has been simulated and is shown in Figure 10. We see that, at the decoded bit-error-rates 10^{-6} and 10^{-10}, the scheme achieves 5.4dB and 7.50dB **real coding gain** over the uncoded QPSK system respectively. The scheme achieves the bit-error probability 10^{-10} at the SNR $E_b/N_0 = 5.50$dB. For SNR $E_b/N_0 = 5.7$dB, the scheme practically achieves **error-free data transmission** .

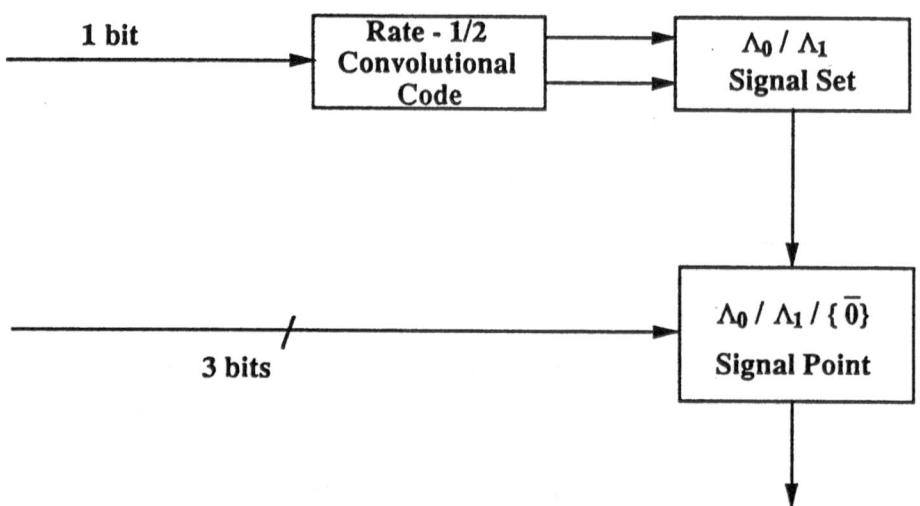

Figure 6 A trellis coded 4-dimensional 8-PSK encoder

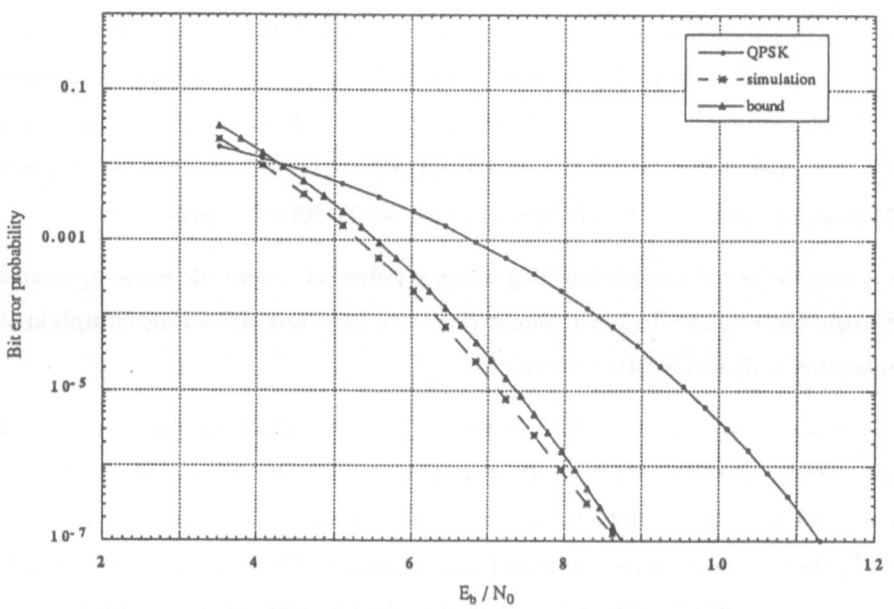

Figure 7   Bit error performance of the 4-state 4-dimensional trellis
8-PSK code

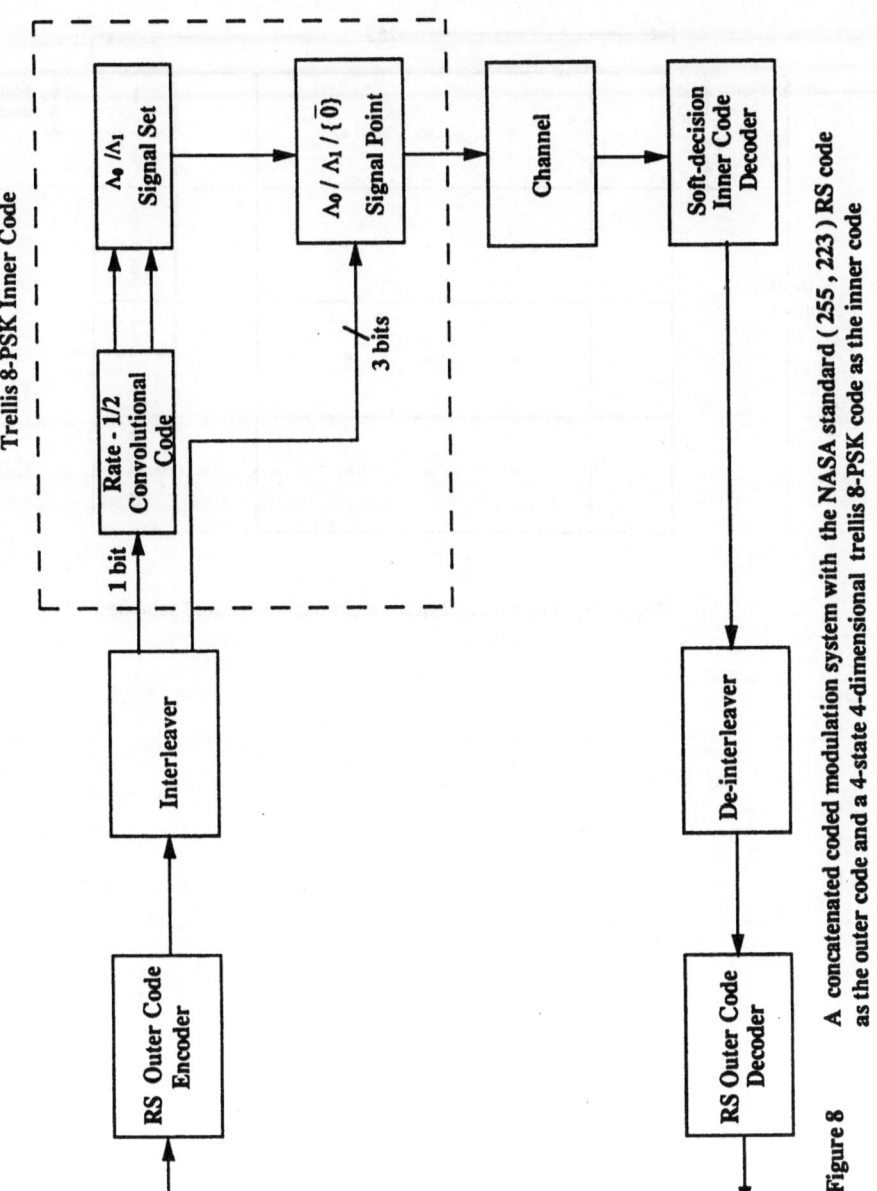

Figure 8   A concatenated coded modulation system with the NASA standard (255 , 223) RS code as the outer code and a 4-state 4-dimensional trellis 8-PSK code as the inner code

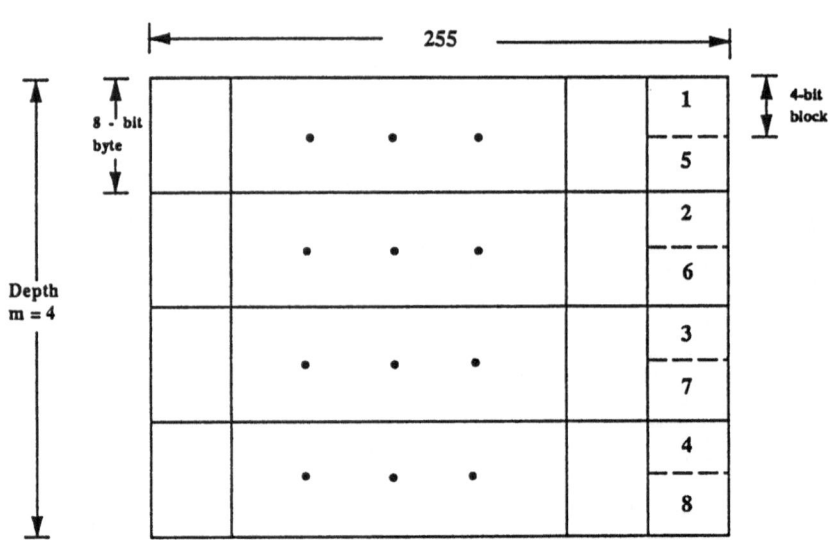

Figure 9   Data array for the second concatenated scheme

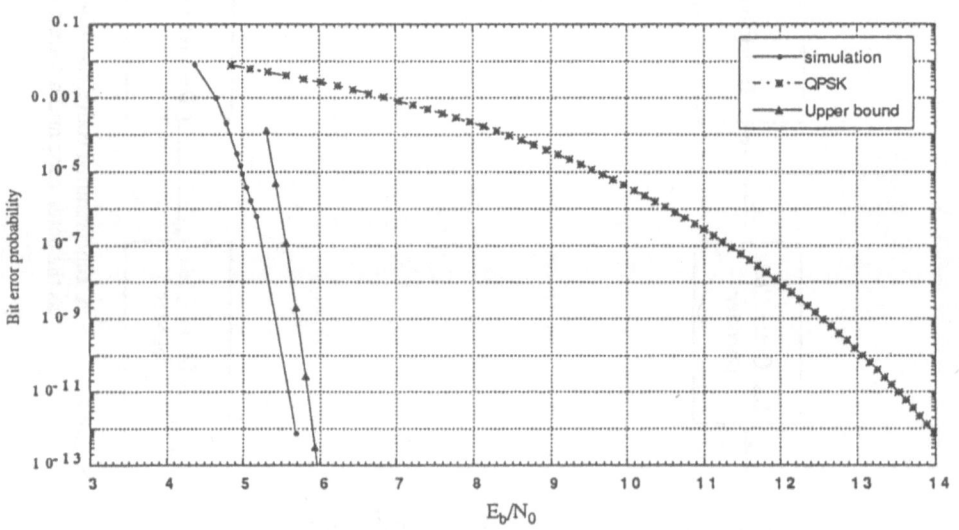

Figure 10   Bit error performance of the second concatenated coded
8-PSK scheme

3. A Two-Level Concatenated Coded Modulation

Using concatenated coded modulation schemes for error control, it is desirable to keep the inner modulation codes **short** (or to keep the number of states of the trellis diagrams for the inner modulation code **small**) to reduce the decoding complexity. However, using a short inner modulation code in a single-level concatenated coded modulation scheme limits the improvement of bandwidth efficiency. This shortcoming can be overcome by using multi-level concatenation and coset inner codes derived from a short modulation code. In fact, multi-level concatenation allow us to to use longer multi-level modulation inner codes with multi-stage suboptimum decoding[9] to achieve very high reliability, large coding gain, and high bandwidth efficiency with reduced decoding complexity.

In this section, we describe a two-level concatenated coded modulation scheme, shown in Figure 11.

Outer Codes

The first-level outer code B_1 is the NASA standard $(255, 223)$ RS code over $GF(2^8)$. The second-level outer code B_2 is the $(255, 239)$ RS code over $GF(2^8)$ with minimum Hamming distance $D_2 = 17$.

Inner Codes

The base modulation inner code Λ_0 is a basic 3-level 8-PSK modulation code of length 8 which is formed from three binary codes, C_{b1}, C_{b2}, for C_{b3}, where : (1) C_{b1} is the (8,1) repetition code with minimum Hamming distance 8; (2) C_{b2} is the (8,7) even-parity code with minimum hamming distance 2; and (3) C_{b3} is the (8,8) universal code with minimum Hamming distance 1. Note that Λ_0 is the same modulation code used as the inner code in the first specific single-level concatenated coded modulation scheme proposed in Section II. Hence, Λ_0 has dimension $k_0 = 16$, minimum squared Euclidean distance $\Delta_0 = 4$, and spectral efficiency $\eta[\Lambda_0] = 2$ bits/signal. The code has a 4-state 8-section trellis diagram as shown in Figure 3.

Let Λ_1 be the basic 3-level 8-PSK modulation code constructed from three binary codes, $C_{b1}^{(1)}, C_{b2}^{(1)}$ and $C_{b3}^{(1)}$, where: (1) $C_{b1}^{(1)}$ is the $(8,0) = \{\bar{0}\}$ code consisting of only the all-zero sequence; (2) $C_{b2}^{(1)}$ is the $(8,1)$ repetition code; and (3) $C_{b3}^{(1)}$ is the $(8,7)$ even-parity check code. Using the signal labeling and bits-to-signal mapping described in Section II, Λ_1 has dimension $k_1 = 8$ and and minimum squared Euclidean distance $\Delta_1 = 8$. Note that $C_{b1}^{(1)} \subset C_{b1}, C_{b2}^{(1)} \subset C_{b2}$, and $C_{b3}^{(1)} \subset C_{b3}$. Hence Λ_1 is a subcode of Λ_0. The coset code $A_1 = \Lambda_0/\Lambda_1$ consists of 2^8 cosets modulo Λ_1, which is used as the first-level inner code.

Let $\Lambda_2 = \{\bar{0}\}$. Then the coset code $A_2 = \Lambda_0/\Lambda_1/\Lambda_2$ consists of 2^8 codewords modulo Λ_2. A_2 is used as the second-level inner code. Both coset codes, A_1 and A_2, have 4-state 8-section trellis diagrams **without parallel transitions** between states, which reduces decoding complexity.

Encoding

Encoding consists of two stages. At the first stage, a code symbol (8 bits) from the first-level outer code encoder selects one of the 64 cosets in the first-level inner code $A_1 = \Lambda_0/\Lambda_1$. At the second stage, a code symbol from the second-level outer code encoder selects one of the 64 signal points (eight 8-PSK signals) from the second-level inner code $A_2 = \Lambda_0/\Lambda_1/\Lambda_2$. Therefore the output of the overall encoder is a sequence of codewords from the 8-PSK modulation code Λ_0. The overall modulation code $\tilde{C}$ has length $255 \times 8 = 2040$, dimension $(223 + 239) \times 8 = 3696$, minimum squared Euclidean distance $D[\tilde{C}] = 132$, and spectral efficiency : $\eta[\tilde{C}] = 3696/2040 = 1.818$ bits/signal.

Decoding

Let $\mathbf{V} = (\mathbf{v}_1, \mathbf{v}_2, \cdots, \mathbf{v}_\ell, \cdots, \mathbf{v}_N)$ be the transmitted code sequence, where $\bar{\mathbf{v}}_\ell$ is a codeword in one of the cosets of the coset code $A_1 = \Lambda_0/\Lambda_1$ and $N = 255$. Let $\mathbf{R} = (\mathbf{r}_1, \mathbf{r}_2, \cdots, \mathbf{r}_\ell, \cdots, \mathbf{r}_N)$ be the received sequence. Decoding is carried out in 2 steps.

At the first-level of decoding, we decode $\mathbf{r}_\ell$ into one of the cosets in $A_1 = \Lambda_0/\Lambda_1$. Based on the decoded coset, we identify the 8-bit byte $\bar{a}_\ell^{(1)}$ which is the estimate of the output byte of the first-level RS outer code C_1 encoder at the time ℓ. Then, decode the sequence $(\bar{a}_1^{(1)}, \bar{a}_2^{(1)}, \cdots, \bar{a}_\ell^{(1)}, \cdots, \bar{a}_N^{(1)})$ based on the RS outer code C_1. Let $(\bar{b}_1^{(1)}, \bar{b}_2^{(1)}, \cdots, \bar{b}_\ell^{(1)}, \cdots, \bar{b}_N^{(1)})$ be the decoded codeword in C_1. Then the input message sequence is retrieved from this decoded codeword. Furthermore, from this de-

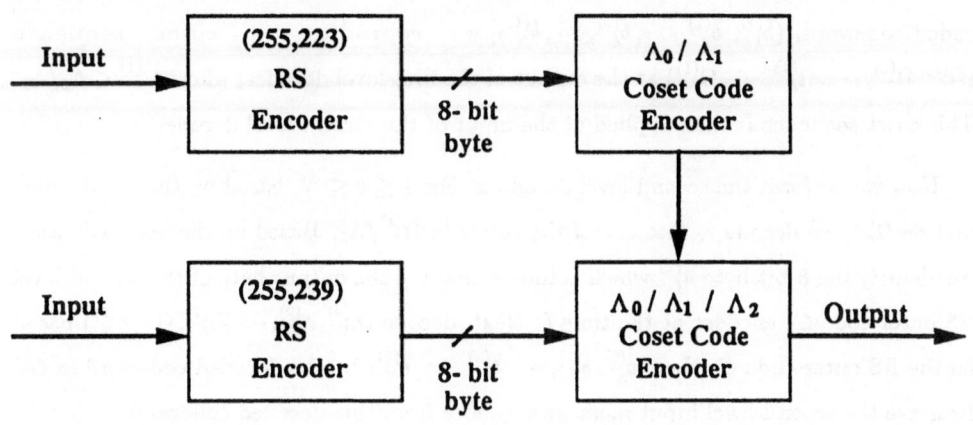

Figure 11 An encoder for the 2-level concatenated coded modulation system

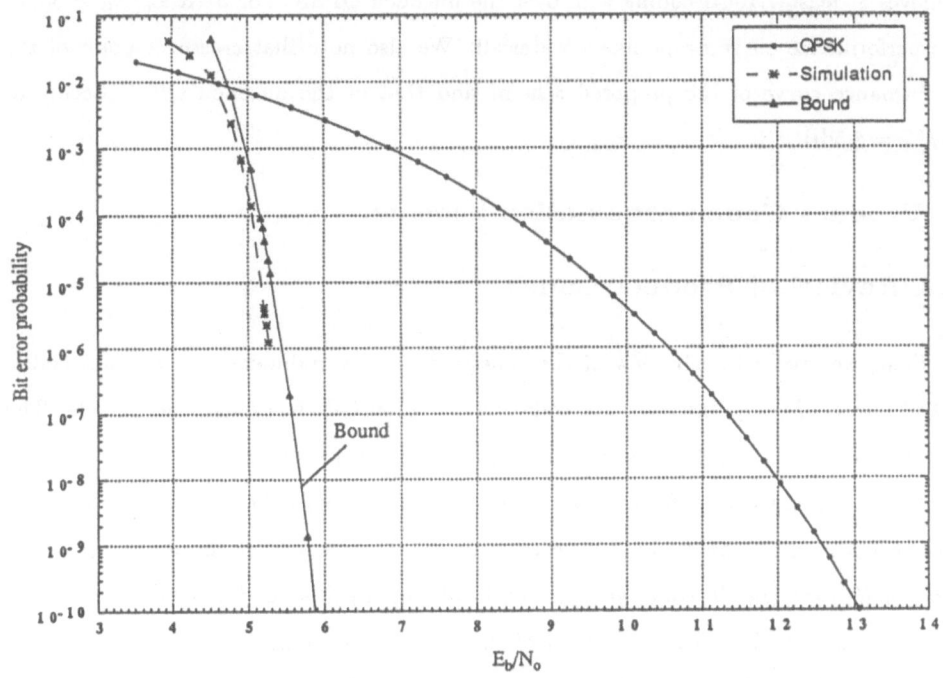

Figure 12 Bit error performance of two-level concatenated coded
modulation scheme

coded codeword, $(\bar{b}_1^{(1)}, \bar{b}_2^{(1)}, \cdots, \bar{b}_\ell^{(1)}, \cdots, \bar{b}_N^{(1)})$, we **reproduce a coset sequence** $(\Omega_1^{(1)}, \Omega_2^{(1)}, \cdots, \Omega_\ell^{(1)}, \cdots, \Omega_N^{(1)})$ at the output of the first-level decoder, where $\Omega_\ell^{(1)} \in \Lambda_0/\Lambda_1$. This coset sequence is then applied at the input of the second-level decoder.

Now we perform the second-level decoding. For $1 \le \ell \le N$, based on the input information $\Omega_\ell^{(1)}$, we decode r_ℓ into one of the cosets in $\Omega_\ell^{(1)}/\Lambda_2$. Based on the decoded coset, we identify the 8-bit byte $\bar{a}_\ell^{(2)}$ which is the estimate of the output byte of the second-level RS outer code C_2 encoder at the time ℓ. Next, decode $(\bar{a}_1^{(2)}, \bar{a}_2^{(2)}, \cdots, \bar{a}_\ell^{(2)}, \cdots, \bar{a}_N^{(2)})$ based on the RS outer code C_2. Let $(\bar{b}_1^{(2)}, \bar{b}_2^{(2)}, \cdots, \bar{b}_\ell^{(2)}, \cdots, \bar{b}_N^{(2)})$ be the decoded codeword in C_2. Retrieve the second-level input message sequence from this decoded codeword.

The performance of this scheme has been analyzed and simulated as shown in Figure 12. We see that the scheme attains very good performance. It achieves a 5.3dB real coding gain over the uncoded QPSK at the bit-error-rate 10^{-6}. At the bit-error-rate 10^{-10}, it achieves at least 7.10dB coding gain over the uncoded QPSK. For SNR $E_b/N_0 \ge 5$dB , the performance curve drops like a waterfall. We also note that crossover point of the performance curve of the proposed scheme and that of the uncoded QPSK occurs at $E_b/N_0 = 4.5$dB.

4. Product Coded Modulation Scheme

4.A Review of Product Codes

First, we give a brief review of the construction of two-dimensional product codes. Let C_1 be an (N_1, K_1) binary block code and C_2 be an (N_2, K_2) binary block code. The product of C_1 and C_2, denoted $C_1 \times C_2$, is formed in three steps. A message of $K_1 K_2$ bits is first arranged in a $K_2 \times K_1$ array of K_2 rows and K_1 columns. Each row of this array is encoded into an N_1-bit codeword in C_1. This row encoding results in a $K_2 \times N_1$ array of K_2 rows and N_1 columns. Then each column of this second array is encoded into an N_2-bit codeword in C_2. As a result of the two-step encoding, we obtain an $N_2 \times N_1$ code array. This code array is then transmitted column by column (or row by row). The collection of all the distinct code arrays form a two-dimensional product code[10]. If the minimum Hamming distances of C_1 and C_2 are d_1 and d_2 respectively, then the minimum

Hamming distance of their product $C_1 \times C_2$ is $d_1 \times d_2$. Henceforth the code C_1 will be loosely termed as the **horizontal code** of the product code and C_2 termed as the **vertical code** of the product code.

4-B. Construction of Product 8-PSK Codes

For $1 \leq i \leq 3$, let $C_{i,1} = (N, k_{i,1}, d_{i,1})$ be a binary block code with minimum Hamming distance $d_{i,1}$ and $C_{i,2} = (n, k_{i,2}, d_{i,2})$ be a binary block code with minimum Hamming distance $d_{i,2}$. Now we form 3 product codes, $P_1 = C_{1,1} \times C_{1,2}$, $P_2 = C_{2,1} \times C_{2,2}$ and $P_3 = C_{3,1} \times C_{3,2}$. Let A, B, and C be three code arrays from P_1, P_2 and P_3 respectively. Let

$$\mathbf{a_j} = (a_{j,1} , a_{j,2} , \cdots , a_{j,n})$$
$$\mathbf{b_j} = (b_{j,1} , b_{j,2} , \cdots , b_{j,n})$$
$$\mathbf{c_j} = (c_{j,1} , c_{j,2} , \cdots , c_{j,n})$$

be the j-th columns of the code arrays, A, B, and C respectively. We form the following sequence :

$$\mathbf{a_j} * \mathbf{b_j} * \mathbf{c_j} = (a_{j,1}b_{j,1}c_{j,1} , a_{j,2}b_{j,2}c_{j,2} , \cdots , a_{j,n}b_{j,n}c_{j,n})$$

For $1 \leq \ell \leq n$, we take $a_{j,\ell}b_{j,\ell}c_{j,\ell}$ as the label for a signal point in the 8-PSK signal constellation as shown in Figure 1. Let $\Lambda(\cdot)$ be the mapping which maps the label $a_{j,\ell}b_{j,\ell}c_{j,\ell}$ into its corresponding signal point $s_{j,\ell}$, i.e., $\lambda(a_{j,\ell}b_{j,\ell}c_{j,\ell}) = s_{j,\ell}$. Then

$$\lambda(\mathbf{a_j} * \mathbf{b_j} * \mathbf{c_j}) = (\lambda(a_{j,1}b_{j,1}c_{j,1}) , \lambda(a_{j,2}b_{j,2}c_{j,2}) , \cdots , \lambda(a_{j,n}b_{j,n}c_{j,n}))$$
$$= (s_{j,1}, s_{j,2}, \cdots , s_{j,n})$$

is a sequence of n 8-PSK signals. For $1 \leq j \leq N$, combining the corresponding columns $\mathbf{a_j}, \mathbf{b_j}$, and $\mathbf{c_j}$ of code arrays A, B and C by the above bits-to-signal mapping, we obtain an $n \times N$ array of 8-PSK signals, denoted $\lambda(A * B * C)$. This array is then transmitted column by column.

Note that, for $1 \leq j \leq N$, the columns $\mathbf{a_j}, \mathbf{b_j}$, and $\mathbf{c_j}$ are codewords from the vertical codes $C_{1,2}, C_{2,2}$ and $C_{3,2}$ respectively. Then, for $1 \leq j \leq N$,

$$\Lambda \overset{\triangle}{=} C_{1,2} * C_{2,2} * C_{3,2}$$

$$= \{\lambda(\mathbf{a_j} * \mathbf{b_j} * \mathbf{c_j}) \; : \; \mathbf{a_j} \in C_{1,2} \, , \; \mathbf{b_j} \in C_{2,2} \text{ and } \mathbf{c_j} \in C_{3,2}\}$$

forms a 3-level 8-PSK modulation code of length n, dimension $k = k_{1,2} + k_{2,2} + k_{3,2}$, and minimum squared Euclidean distance $D[\Lambda] = \min\{0.586 \times d_{1,2} \, , \; 2 \times d_{2,2} \, , \; 4 \times d_{3,2}\}$. Consequently, the following collection of distinct arrays of 8-PSK signals,

$$\Omega = \{\lambda(A * B * C) \; : \; A \in P_1 \, , \; B \in P_2 \text{ and } C \in P_3\}$$

form a **product** 8-PSK modulation code of length nN, dimension $k_{1,1} \times k_{1,2} + k_{2,1} \times k_{2,2} + k_{3,1} \times k_{3,2}$, and minimum squared Euclidean distance

$$D[\Omega] = \min\{0.586 \times d_{1,1} \times d_{1,2} \, , \; 2 \times d_{2,1} \times d_{2,2} \, , \; 4 \times d_{3,1} \times d_{3,2}\} \tag{1}$$

From (1), we see that we can construct product 8-PSK codes with **arbitrarily** large minimum squared Euclidean distance by choosing the horizontal and vertical component codes properly. The overall encoder for a product 8-PSK modulation code is shown in Figure 13.

4.C Decoding

One obvious way (though impractical) of decoding the proposed product modulation code is to compare each received code sequence with all the possible code sequences and find the closest one in terms of minimum squared Euclidean distance. The decoding complexity associated with this technique would be simply enormous. We will focus on a suboptimal decoding procedure which allows decoding of the codes with reduced decoding complexity while maintaining good performance. Recall that each codeword in Ω can be written in the form $\mathbf{V} = (\mathbf{v_1}, \mathbf{v_2}, \cdots, \mathbf{v_N})$ with $\mathbf{v}_i \in \Lambda$ for $1 \leq i \leq N$. Let the received sequence be $(\mathbf{r_1}, \mathbf{r_2}, \cdots, \mathbf{r_N})$. The decoding is performed in 3 stages. At the first stage of decoding, P_1 is decoded, followed by P_2 and lastly P_3. At the first stage of decoding, each of the columns of P_1 is decoded, using a multi-stage soft-decision decoding algorithm for Λ, based on the received sequence. Recall that the columns of P_1 are codewords in $C_{1,2}$ which are associated with the first labeling level of Λ. The horizontal code $C_{1,1}$ of P_1 uses these decoded estimates of the columns to decode the horizontal rows of bits. The horizontal row decoding is advantageous since it corrects additional errors in the columns and thus helps reduce the error propagation into the next stage of decoding. In fact, if the horizontal codes are chosen powerful enough, the error propagation effect is negligible, thereby eliminating the major problem of multi-stage decoding and improving reliability tremendously. Thus after the horizontal row decoding of P_1 we have new estimates of the columns (the corrected estimates). These new estimates along with the received sequence are passed to the second decoding stage and are used to decode the columns of P_2. The decoding of P_2 and P_3 follows the same pattern as that of P_1.

4.D A specific Product coded 8-PSK scheme

Let $C_{1,2}$ be the $(16, 5, 8)$ Reed-Muller(RM) code with minimum distance 8 , $C_{2,2}$ be the $(16, 15, 2)$ even parity code with minimum distance 2, and $C_{3,2}$ be the trivial $(16, 16, 1)$ code with minimum distance 1. These three vertical codes are used to form Λ. Also let $C_{1,1}$ be the sixteen-error-correcting $(1023, 863, 33)$ BCH code, $C_{2,1}$ be the $(1023, 943, 17)$ BCH code with error correcting capability $t = 8$ and $C_{3,1}$ be the triple-error-correcting $(1023, 993, 7)$ BCH code. Then the spectral efficiency of the product 8-PSK code Ω is $\eta = 2.098484$ bits/symbol and the phase invariance is $45°$. Figure 14 shows the upper bound on the probability of bit error. The bound on the bit error rate of Ω has been calculated assuming a 3-stage soft-decision decoding for Λ. The $(16, 5, 8)$ RM code associated with the first labeling level of Λ has a 8-state trellis which is used to decode the columns of P_1 at the first stage of decoding with the Viterbi algorithm. Also, the $(16, 15, 2)$ even parity code associated with second level of Λ has a very simple 2-state trellis which is used to form the decoded estimates of the columns of P_2 (i.e. the second stage of decoding). The decoding complexity associated with decoding the third level of Λ is very small and hence will be ignored. The total decoding complexity of Λ due to the soft-decision multi-stage decoding is therefore $8 + 2 = 10$ states. As can be seen from Figure 14, the suboptimum decoding allows one to achieve at least 3.75dB coding gain at the bit error rate of 10^{-6} and 6.0dB coding gain at 10^{-10} bit error rate over the uncoded QPSK. The decoding complexity of the horizontal codes of the product codes depends upon what kind of decoding algorithm is chosen for the BCH codes.

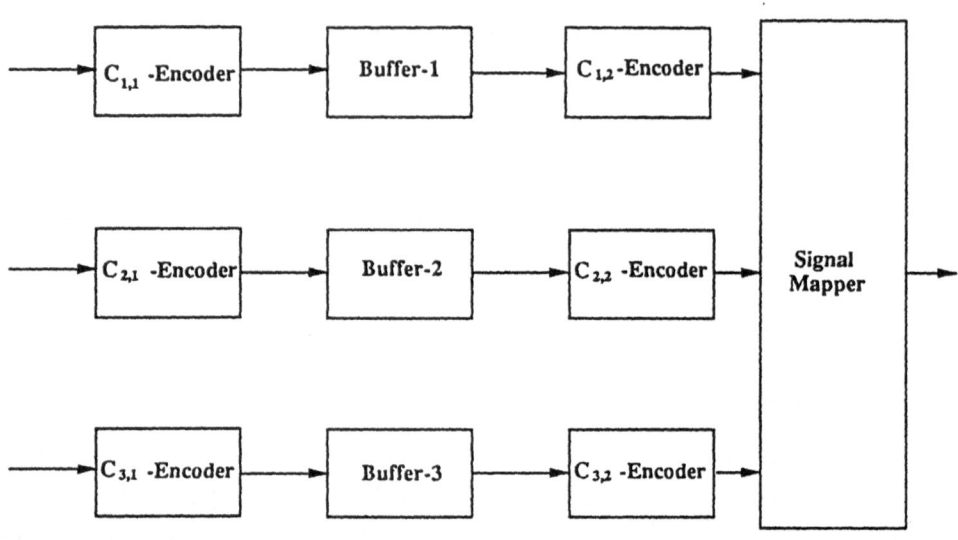

Figure 13 Overall encoder for a product 3-level 8-PSK modulation code

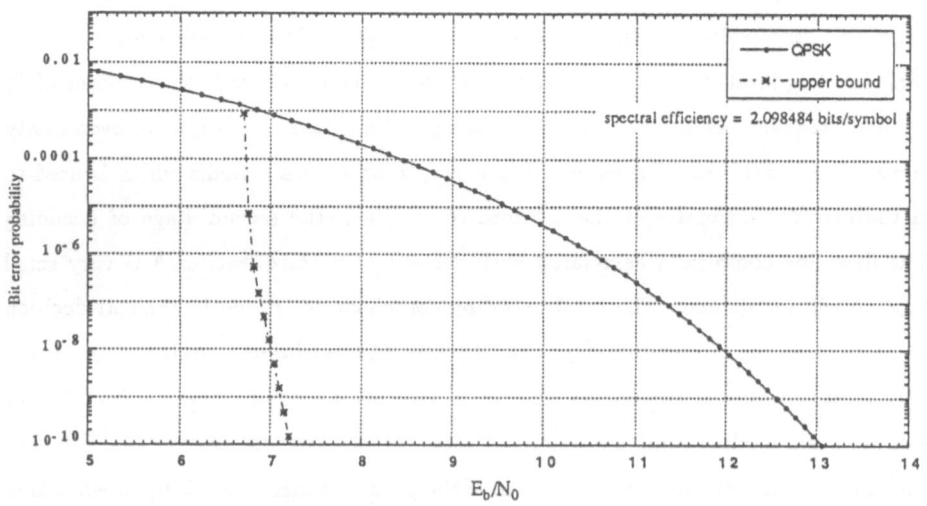

Figure 14   Upper bound on the bit error probability for the product
coded scheme

REFERENCES

1. G. Ungerboeck, "Channel Coding with Multilevel/Phase Signals," **IEEE Trans. on Information Theory**, Vol. IT-28, No. 1, pp. 55-67, January 1982.

2. G.D. Forney, Jr., **Concatenated Codes**, MIT Press, MA, 1966.

3. T. Kasami, T. Takata and T. Fujiwara and S. Lin, "A Concatenated Coded Modulation Scheme for Error Control," **IEEE Transactions on Communications**, Vol. 38, No. 6, June 1990.

4. H. Imai and S. Hirakawa, "A New Multilevel Coding Method Using Error Correcting Codes," **IEEE Trans. on Information Theory**, Vol. IT-23, No. 3, pp. 371-376, May 1977.

5. V.V. Ginzburg, "Multidimensional Signals for a Continuous Channel," **Problemy Peredachi Informatsii**, Vol. 20, No. 1, pp. 28-46, 1984.

6. S.I. Sayegh, "A Class of Optimum Block Codes in Signal Space," **IEEE Trans. on Communications**, Vol. COM-30, No. 10, pp. 1043-1045, October 1986.

7. T. Kasami, T. Takata and T. Fujiwara and S. Lin, "On Multi-Level Block Modulation Codes," **IEEE Trans. on Information Theory**, Vol. IT-37, No. 4, July 1991.

8. S. Rajpal, D.J. Rhee and S. Lin, " Multidimensional MPSK Trellis Codes," **Proceedings of the 14th Symposium on Information Theory and its Applications** , Ibusuki, Japan, December 11-14, 1991.

9. T. Kasami, T. Takata and T. Fujiwara and S. Lin, "Multi-Stage Decoding for Multilevel Block Modulation Codes," NASA Technical Report No. 91-002 (August 27, 1991) and submitted to **IEEE Trans.** on Information Theory, 1991(in revision).

10. S. Lin and D.J. Costello, Jr., **Error Control Coding: Fundamentals and Applications**, Prentice-Hall, Englewood Cliffs, New Jersey, 1983.

Synchronization Aspects of a Mobile Satellite Voice and Data Modem

B. Koblents, P.J. McLane and W. Choy

Department of Electrical Engineering

Queen's University, Kingston, ON, K7L 3N6

Abstract

We present results for an asynchronous timing recovery algorithm for three mobile satellite modems operating at either 2400 bps or 4800 bps. Both coherent and differentially coherent modems are considered. The timing recovery is completely based on choosing sample points for detection and timing control based on an oversampled and interpolated input signal. As such the sole modem interface is a uncontrolled A/D convertor and timing recovery is asynchronous with the receiver A/D clock. In addition, a polyphase filter interpolation timing recovery algorithm is modified and tested. The satellite modem applications involve filtered BPSK, filtered QPSK and 8–PSK with trellis coded modulation.[*]

1 Introduction

The first application we consider is a 6000 bps aeronautical/satellite modem based on Raised Cosine (RC) filtered phase coherent BPSK modulation, with 40% excess bandwidth. To permit the number of FIR taps in the receiver data filters to be kept low, we have developed a successful modem concept using polyphase filter interpolation for timing recovery. The modem receiver recovers the timing phase from the oversampled input signal, leaving the input A/D converter free running, with no changes possible in it's timing phase. Such timing recovery techniques for digitally implemented modems are discussed in [1]. Two other modem families are discussed in the paper, both intended for voice rate satellite applications: rectangular pulse shaped modems, one employing uncoded DQPSK, and another with trellis coded 8–DPSK, both running at 2400 bps; and a family of RC shaped DQPSK modems operating at 4800 bps, with excess bandwidth factors ranging from 100% to 60%. The latter two families also employ asynchronous timing

---

[*] This research was done at Queen's University and was sponsored by the Telecommunications Research Institute of Ontario (TRIO).

recovery, but do not require polyphase interpolation for timing correction, due to a higher degree of oversampling.

2 Polyphase Filter Interpolated Timing Recovery

Figure 1 shows the block diagram for the 6000 bps RC filtered, coherent, BPSK modem which employed four samples per symbol. Upon start-up, the modem receiver is loaded with data filters F_1, which assume zero timing offset. Suppose there is an offset in the timing phase, such that the impulse response of the raised cosine pulse shape exhibits intersymbol interference (ISI) and is as shown in Figure 2. The timing error detector detects the degree of ISI, by estimating the impulse response samples h_1 and h_{-1} in Figure 2, and produces an error signal large enough to require timing correction. Instead of adjusting the input A/D so that the impulse response returns to it's unperturbed shape, it is possible [1], to switch filter coefficients to those of F_2, so that the receiver filters are once more matched to the sampling offset perturbed transmit filter for the incoming waveform. With a receiver running asynchronously with the transmitter, a small difference in sampling frequencies between them is almost inevitable, hence the timing offset will be continuously growing. When the timing error signal again exceeds the correction threshold, a new set of filter coefficients, F_3 is loaded. The procedure is repeated until the timing perturbation reaches the next sample of the original response. Then, the filters are reset to F_1, and a sample is either added, or dropped. Because the new filter responses lie in between the sample points of the original response, they can be obtained by interpolation between the samples of the original response. The interpolation can be made as fine grained as desired, by simply implementing the receiver filters oversampled to the desired interpolation factor, M, and then obtaining M individual interpolating filters, by taking successive Mth taps out of the oversampled response. These turn out to the branches of M−branch polyphase filter decomposition of the oversampled interpolating filter [1]. The polyphase filter approach is simply realized in DSP. The interpolating FIR filter coefficients are simply stored in memory and loaded when needed.

The timing error detector used in the 6000 bps modem is one proposed by Mueller and Mueller [2]. We refer to it as the M&M algorithm. The idea is that the error signal can be expressed as in (1), as shown in Figure 2. As in [2] and [19], the timing error signal is, ideally

$$e = h_1 - h_{-1} \,. \tag{1}$$

Here h_i is the impulse response of the in-phase filter in Figure 1 which is the data detection filter. Also, e is illustrated in Figure 2. It can be shown that an unbiased estimate (see the Appendix) of the timing error can be obtained from

$$\hat{e}_i = \hat{a}_{i-1}\, r_i - \hat{a}_i\, r_{i-1} \tag{2}$$

which is the algorithm implemented in Figure 1. Note that it is decision directed (DD), requiring estimates of the current and previous bits to function. The timing phase is updated according to

$$\tau_{k+1} = \tau_k + \Delta e_k \qquad (3)$$

where Δ is the step size and τ is the timing phase. When τ_k crosses a threshold either a new filter is added or a sample is dropped or added where we determine the threshold by trial and error. A key point on the rule in (3) is that the error signal does not depend on the response peak of the current symbol, h_0, resulting in a relatively low value of pattern jitter [1]. Some other properties of the M&M algorithm can be found in the Appendix of this paper. We use Gardner's zero-crossing timing recovery algorithm [3] to set-up the modem. Fifty symbols were allowed for modem training. The approach in [3] will be described later in the paper. Also, coherent operation is obtained in Figure 1 through a Costas loop. The decoupling of the timing and phase loops is treated in [19]. In particular, we employ a simple algorithm that avoids phase lock loop hang-up [4].

3 BER Tests with 6000 bps Modem

The 6000 bps modem and an additive white Gaussian noise link have been simulated using Turbo Pascal, and the receiver tested. The results are shown in Figure 3 for 16 tap FIR square root RC filters. A timing offset of $T/32$ is introduced every 200 or 400 symbols, corresponding to timing offsets of 0.5 Hz and 1 Hz respectively. The error rate range of interest for mobile satellite voice applications is 10^{-1} to 10^{-3}, corresponding effective SNR of 2 to 4 dB in E_b/N_o. As seen in Figure 3, the simulation predicts relatively good performance for signal to noise ratios as low as 1 dB. Due to the fact that the timing error detector in (2) is decision directed, the performance degrades for error rates greater than 10^{-1}. Note that the simulated rates of timing drift are much higher than those observed in systems implemented on recent generations of DSP boards. The system is currently being implemented on Texas Instruments TMS32025 signal processors.

4 TCM Modem

Our next application involved a trellis coded M-DPSK modem which was also developed in [10]. The modem is aimed at MSAT 4800 bps digital speech service. Herein we describe a MSAT modem that is trellis coded and is realized on two second generation DSP chips. The modem operates at 2400 bps and is unfiltered in that rectangular pulse shapes are used.

A block diagram of the communications link is given in Figure 4. Also shown are the DSP chips used to represent portions of the overall system. The design of the components in the transmitter and receiver are based on theoretical studies in [8,9]. The trellis coder used is Ungerboeck's 8–state trellis code [7]. The experimental results to be shown later are based on 2400 bps operation using a TMS 32010 for the Transmitter and Receiver and a TMS 320C25 for the Viterbi decoder and convolutional deinterleaver. An earlier treatment of our TCM modem design can be found in [5].

5 TCM Transmitter

The block diagram of the all digital transmitter that uses a 4–kHz carrier is shown in Figure 5. We use approximately 8 samples per cycle in our digital implementation. An analog transmit filter smooths the 4–kHz carrier and provides interpolation between the eight carrier sample points. The algorithm to generate the carrier is taken from [11]. The interleaver and trellis encoder are described in [8,9]. Currently rectangular pulse shapes are used. The trellis coder and convolutional interleaver used in our experiments are given in [9]. One reason for using a convolutional interleaver is its similar structure to the convolutional encoder used in the trellis encoder. Also it is twice as efficient [9] for the same delay relative to the block interleaver used in [10].

The differential eight phase modulator is based on the equation

$$\theta_{k+1,m} = (\theta_{k,m} + \lambda_k) \, mod \, 2\pi \tag{4}$$

where λ_k is the information phase. The phase above in the $(k+1)st$ modulation interval is used to discretely phase modulate the digitally generated carrier. This is done by initializing the 8–sample point per cycle carrier from one of eight sample points. Note that no multiplication is involved as would occur in a QAM-based realization. This completes the description of the Transmitter in Figure 5.

6 TCM Receiver

The receiver block diagram is also shown in Figure 5. The input to the receiver is the differentially phase modulated, 4.8 kHz carrier plus noise. This is converted at a rate of 38.4 k samples/sec to a sampled waveform by the front-end, A/D convertor. The numerical oscillator (NO) box in the receiver generates the free running, digital sinusoid of reference [11]. In-phase and quadrature components of the received waveforms then form the inputs to the data and timing integrators. We will cover our timing algorithm later. We now consider the metric computation and Doppler phasor correction process.

6.1 Detection and Doppler Correction

Our Doppler correction algorithm is discussed in a companion paper [18]. We use open-loop Doppler correction [12] in the set-up mode and the tracking technique of [7,13,14] for the data mode. Stream data transmission is assumed. The method was tested in simulation in [18] for the shadowed fading scenarios in [17]. We are presently incorporating it into our experimental modem. In our approach we correct the Doppler phasor rather than the Doppler frequency as it is easier to implement in DSP realizations.

To derive our Doppler recovery algorithm let us regard the in-phase and quadrature as distortionless. That is, the double frequency terms have been completely removed through averaging in the integrators. Also, noise will be ignored in our derivation but is considered in [18]. Then the I and Q sampled signals from the data integrators in Figure 5 can be written as

$$Z_k = I_k + j \, Q_k \tag{5}$$

$$Z_k = A_k \, e^{j\theta_{k,m}} \, e^{jkw_DT} \, e^{j\theta} \tag{6}$$

where

$$
\begin{aligned}
A_k &= k\text{th amplitude sample} \\
\theta_{k,m} &= \text{modulation phase in (1)} \\
w_DT &= \text{Doppler angle} \\
\theta &= \text{offset phase (slowly varying)} \, .
\end{aligned}
$$

The differentially detected signal is

$$\Gamma_k = Z_k \, Z_{k-1}^* \tag{7}$$

where Z_{k-1}^* is the complex conjugate of Z_{k-1}. Now use of (6) yields

$$\Gamma_k = A_k \, A_{k-1} \, e^{j\lambda_k} \, e^{jw_DT} \tag{8}$$

where λ_k is the data phase in (1). Note that the Doppler phasor e^{jw_DT} perturbs Γ_k. If we remove it via the operation, $\Gamma_k \left(e^{jw_DT}\right)^*$, where $e^{\widehat{jw_DT}}$ is the estimate of the Doppler phasor, the detection of λ_k is

$$\hat{\lambda}_k = arg\left(\Gamma_k \, e^{-\widehat{jw_DT}}\right) \tag{9}$$

for the uncoded case. For the trellis coded case the signal $\Gamma_k \, e^{-\widehat{jw_DT}}$ forms the input to the Viterbi, trellis decoding algorithm.

6.2 Timing Recovery in Rectangular Pulse Shaped Systems

The timing error detector used is a variation of Gardner's zero crossing tracker [1,3,15] and uses mid-pulse to mid-pulse, or synchronization integrators as in Figure 5. The scheme is based on the idea that a mid-pulse to mid-pulse integration should be zero through a data transition. The timing error tracker output depends on the synchronization as well as the data integrators on both rails, as follows:

$$E_k = (I_k - I_{k-1}) I_{k-\frac{1}{2}} + (Q_k - Q_{k-1}) Q_{k-\frac{1}{2}} \tag{10}$$

where I_k, Q_k are the data integrator outputs, and the $I_{k-\frac{1}{2}}$ and $Q_{k-\frac{1}{2}}$ are the synchronization integrator outputs in Figure 5. It is relatively easy to show that the tracking term is invariant to phase offsets [1,3]. This is important because our method of detection is differential detection and this can follow the timing recovery process. Unlike the M&M tracker presented earlier, this tracker is non-decision-aided (NDA), although it is possible to build a DD system built on the same principles [1,3]. The advantage of the NDA implementation is that it is free of self noise for any M-PSK modulation, and thus is not restricted to signal constellations with even pulse heights [20].

In the absence of timing offset, the double frequency terms produced in the mixers are completely averaged out in the integrators. With a timing offset, however, these double frequency terms are not fully averaged out, and can cause false error readings on data transitions not involving 180° phase shifts [13]. The timing jitter caused by the double frequency distortions was adversely affecting the BER performance of the modems, as shown in Figure 6. Simulation studies showed that by taking the sum of the absolute value of the synch accumulators, and only proceeding to make timing measurements if the sum did not exceed a certain threshold as in (11), the timing jitter could be significantly reduced. Our algorithm is

$$\text{ABSUM} = \text{ABS}\left(I_{k-\frac{1}{2}}\right) + \text{ABS}\left(Q_{k-\frac{1}{2}}\right)$$
$$\text{if ABSUM} \leq \text{THRESHOLD then} \tag{11}$$
$$\text{make timing measurements .}$$

A summary of the simulation results averaged over the SNR range of interest (9 to 20 dB, $2E_b/N_o$) is shown in Table 1. A threshold level of 60% of the maximum data accumulator value caused timing decisions to be skipped for 60% of the received symbols, and reduced the number of decisions on pulses vulnerable to double frequency interference by 67%. Making the threshold tighter caused the algorithm to miss larger numbers of correction on the desired 180° phase shifts. Similar behaviour was observed when the thresholding scheme was implemented on the DSP system, and the timing jitter was reduced with an attendant improvement in performance.

The error signal is then filtered with a 300 Hz first order IIR loop filter, and applied to the timing correction mechanism. If the error signal exceeds a threshold for timing correction, then the modem either drops or adds a sample to the next symbol's integration period. Since the system is highly oversampled, there is no need for polyphase filter coefficient interpolation. Note that once more, timing is corrected without altering the phase of the input A/D converter, which remains free running.

6.3 Deinterleaving and Decoding

The deinterleaver and Viterbi decoder are implemented on a second generation signal processor. This is as shown in Figure 5. This block can operate at 4800 bps. The decoding algorithm that was implemented was Ungerboeck's modified Viterbi algorithm [16]. As there is no intersymbol interference and PSK signalling is assumed. The metric is $Re(\alpha_n^* Z_n)$ where Z_n is the complex integrator output and $\alpha_n = \lambda_n$, the assumed, trellis encoded, PSK symbol.

7 Results: Rectangular Pulse Shaped Modems

The rectangular pulse shaped modems were tested over a Gaussian noise channel. Our first experimental result is for the 2400 bps uncoded DQPSK system. A 100 symbol set up period was used established receiver timing. After that a long sequence of pseudo random symbols was transmitted and the errors counted. The results of the experiment are given in Figure 6, which shows three curves: the performance of the system before any changes were introduced; the performance when the thresholding scheme of (11) is included; and a computer simulation with perfect timing, which serves as the ideal case. Note that the thresholding scheme reduces implementation losses by roughly 1/2 dB, to roughly 1 dB in E_b/N_o. Experimental results for the 8–DPSK trellis coded modem incorporating the thresholding of (11) are shown in Figure 7. With proper choice of loop filter implementation losses are reduced to a small fraction of a dB. Note that implementation losses are lower in the coded case. This is due to the error correction capability of this system.

8 RC Filtered DQPSK Modem

The MSAT signal standard will use 60% excess bandwidth (α) RC pulse shapes and DQPSK modulation with conventional convolutional coding. As such, a family of 4800 bps modems has been developed based on the concepts presented earlier. The transmitter and receiver block diagrams are shown in Figure 8. Doppler correction is not shown as it has been incorporated in our modem. Gardner's timing algorithm of equation (10) is used, except this time there is no need to use separate synchronization filters, as the mid points of a symmetric pulse shaped response are zero crossings, at least in the 100% α case. If the α is less than 100%, the mid points of the response are not entirely free of ISI, and the interference gets worse as α is reduced. Still, with properly tight loop filtering, the ISI induced self noise can be controlled sufficiently so that performance is not significantly degraded [20]. The thresholding scheme employed in the rectangular pulse modems is not needed here, since the data filters attenuate the double frequency terms from the mixers. Polyphase filter interpolation timing correction has been implemented in simulation. Figure 9 shows timing error in the noise free simulation case without filter interpolation. Figure 10 shows the same perturbation sequence, but this time with filter coefficient interpolation using an interpolation factor of four. Note that the timing error is kept much smaller if interpolation is used. For initial implementation on the DSP boards, asynchronous timing correction was by dropping and adding samples only, as in the rectangular pulse shaped systems. Preliminary BER results without the interpolation are good enough to not require any additional work. The reason is that with 8 samples per symbol, the impulse response samples are closely enough spaced not to require any interpolation. Figure 11 shows the experimental BER result for the 100% excess bandwidth case. The results for the 80% and 60% cases are given in Figures 12 and 13 respectively. Note that the implementation losses are low.

9 Speech Coder Application

The 4800 bps, 60% RC modem just described has been interfaced to a 4800 bps LPC speech coder/decoder provided by SkyWave Electronics of Kanata, Ontario [21]. The major challenge in this experiment was the synchronization of our asynchronous timing recovery algorithm in our modem to a separate circuit board containing the speech coder chip with its own clock. Fortunately, timing corrections occur infrequently in our 4800 bps modem. Adding or dropping a sample represents a change in duty cycle that must be synchronized to the speech coder/decoder system. When 16 samples per symbol were used a sample added or dropped represents a change in duty cycle of 1/16th of a symbol interval. No loss in synchronization occurred in this case, but this represents a modem speed of only 2400 bps. The 4800 bps case requires 8 samples per symbol interval. Our speech coder/decoder synchronization algorithm in this case just divided the timing correction into two steps so as to mimic the case of 16 samples per symbol interval. This combined approach has worked well and the modem gives no apparent perceptual change to the quality of the speech coder/decoder. We also reduced the number of timing corrections in the timing recovery algorithm to ease the speech coder/modem synchronization problem. This resulted in a highly stable operation with no perceptual insertion loss due to the modem.

10 Conclusions

Timing recovery in either coherent or differentially coherent mobile satellite modems can be based on asynchronous techniques, resulting in simple, robust timing error correctors, and removing any need to adjust the timing phase of the receiver A/D converters. Either the Gardner [1,3], or the Mueller and Mueller [2] timing error detectors can be used, with the Gardner technique applicable to PSK signal constellations with uneven pulse heights. The M&M timing error detector remains useful for α of less than 50%, while α of 60% or more permit the use of the Gardner technique. For systems with a limited number of samples per symbol, polyphase filter coefficient interpolation can be used to improve performance. Finally, the modem has been shown to operate in either digital speech or data modes.

Acknowledgments

Mike Belanger, Ed Wiseman and Tuck Tay of Queen's University are thanked for their help in the project.

References

[1] F.M. Gardner, "Demodulator Reference Recovery Techniques Suited for Digital Implementation", ESA, Final Report, May 16, 1988.

[2] K.H. Mueller and M. Mueller, "Timing Recovery in Digital Synchronous Data Receivers", IEEE Trans. Comm., COM-24, pp. 516–531, May 1976.

[3] F.M. Gardner, "A BPSK/QPSK Timing-Error Detector for Sampled Receivers", IEEE Trans. Comm., Vol 34, pp.423–429, May 1986.

[4] F.M. Gardner, "Hang-up in Phase Lock Loops", IEEE Trans. Comm., Vol. COM-25, pp. 1210–1214, Oct. 1977.

[5] P. Bengough, L. Berg, B. Koblents, P.J. McLane, "Signal Processor Based MSAT Trellis Coded M-DPSK With Adaptive Doppler Phasor Correction", Canadian Conf. Computer and Elec. Eng., Ottawa Ont., Sept. 1990.

[6] A. Papoulis, *Probability, Random Variables and Stochastic Processes*, 2nd Edition, McGraw-Hill, New York, 1984, page 188.

[7] G. Ungerboeck, "Channel Coding with Multilevel/Phase Signals", *IEEE Trans. on Information Theory*, Vol. IT-28, pp. 56–66, January 1982.

[8] P.J. McLane, P.H. Wittke, P. Ho and C. Loo, "PSK and DPSK Trellis Codes for Fast Fading Shadowed Mobile Satellite Communication Channels", *IEEE Trans. on Comm.*, Vol. 36, pp. 1242–1246, November 1988

[9]   A. Lee and P.J. McLane, "Convolutionally Interleaved PSK and DPSK Trellis Codes for Shadowed, Fast Fading Mobile Satellite Communication Channels", *Vehicular Technology Conference*, Philadelphia, PA, June 1988: see also, *IEEE Trans. Veh. Tech.*, 1990.

[10] T.C. Jedrey, N.E. Lay and W. Rafferty, "An 8–DPSK TCM Modem for MSAT-X", *Proc. Mobile Satellite Communications Conference*, JPL, Pasadena, Ca., May 3–5, 1988.

[11] K-S Lin, editor, *Digital Signal Processing Applications with the TMS 320 Family*. Volume 1, Part II, Article #8 by D. Garcia, Prentice-Hall, 1987, Englewood Cliffs, NJ.

[12] M.K. Simon and D. Divsalar, "Doppler-Corrected Differential Detection of MPSK", *IEEE Trans. Comm.*, Vol. 37, pp. 99–110, February 1989.

[13] F. Edbauer, "Interleaver Design for Trellis-Coded Differential 8–PSK Modulation with Non-Coherent Detection", *Proc. Mobile Satellite Conf.*, JPL, Pasadena, Ca., May 3–5, 1988: see also, *IEEE Journal of Selected Areas of Communication*, August 1989.

[14] F.D. Natali, "Noise Performance of a Cross-Product AFC with Decision Feedback of DPSK Signals", *IEEE Trans. Communications*, Vol. COM34, pp. 303–307, March 1986.

[15] S.J. Henely, "Modem for the Land Mobile Satellite Channel", *Proc. Mobile Satellite Conference*, JPL, Pasadena, Ca., May 3–5, 1988.

[16] G. Ungerboeck, "Adaptive Maximum Likelihood Receiver for Carrier-Modulated Data Transmission Systems", *IEEE Trans. Comm.*, Vol. 22, pp. 624–636, May 1974.

[17] C. Loo, "A Statistical Model for a Land Mobile Satellite Link", *IEEE Trans. Vehicular Technology*, Vol. VT-34, pp. 122–127, August 1985.

[18] P.J. McLane, "Two-Stage Doppler Phasor Corrected TCM/ DMPSK for Shadowed Mobile Satellite Channels", *Canadian Conference on Electrical and Computer Engineering*, Ottawa, September 4–6, 1990, to appear, *IEEE Trans. Comm.*

[19] P.J. McLane, W. Choy and T. Tay, "Filter Coefficient Interpolated Timing Recovery in Sampled Coherent PSK Receivers", submitted to *Globecom'92*.

[20] B. Koblents, *Asynchronous Timing Recovery in M-PSK Data Modems*. M.Sc. thesis in preparation, Queen's University, Kingston, Ontario.

[21] P. Rossiter, private communication, SkyWave Electronics, Kanata, Ontario, Canada.

Appendix A

Herein we derive some properties of the M&M timing recovery system from [2]. Recall that for a type A system from [2]

$$e = h_1 - h_{-1}$$

where h_1 and h_{-1} are shown in Figure 2. An estimate of e is

$$\hat{e}_i = r_i\,\hat{a}_{i-1} - r_{i-1}\,\hat{a}_i \qquad (A\text{-}1)$$

where $\hat{a}_i = \text{sgn}(r_i)$, a data decision for binary data and the detector input is

$$r_i = \sum_k a_k\,h_{i-k} + n_i$$

with n_i independent system noise. Here the h_i are samples of the overall RC pulse shown in Figure 2.

A-1 Unbased Estimate of e

If we assume $\hat{a}_i = a_i$, that is no detection errors, then as the a_i are independent and $+/-1$, from (A-1) we have

$$\langle \hat{e}_i \rangle = \langle r_i\,a_{i-1} \rangle - \langle r_{i-1}\,a_i \rangle$$
$$= h_1 - h_{-1} = e\,.$$

where $<x>$ is the statistical average of x. This result shows that $\hat{e}_i$ is an unbiased estimate of e [6]. A plot of e for $\sqrt{RC}\ T_x$ and R_x filters and $\alpha = 0.4$ is given in Figure 12. The timing correction function e is linear for $0 \le |\tau| \le T/4$ which is highly desirable.

A-2 Pattern Jitter

The variance of $\hat{e}_i$ is known as the pattern jitter. A low pattern jitter is required for the timing error loop in (3) to settle rapidly.

For error free conditions, one can show that

$$\sigma^2(\tau) = v_{ar}(\hat{e}_i - e)$$
$$= 2 \sum_{\substack{k=-N \\ k \neq 0,1,-1}}^{N} h_k^2 + h_{-1}^2 + h_1^2 + 2\sigma^2 \qquad (A\text{-}2)$$

where var(x) is the variance of the random variable x, the extent of the ISI is from $-NT$ to NT and $\sigma^2 = E(n_i^2)$. A key point is that h_o, the largest sample of h_k, cancels in the Type A rule (2) from [2]. This leads to low pattern jitter in (A-2).

A Type B rule is $e = h_1$ or $\hat{e}_i = \hat{a}_{i-1}\, r_i$. This is also unbiased but contains h_o and thus has huge pattern jitter, $viz,$

$$\sigma^2(\tau) = h_o^2 + \sum_{\substack{k \\ k \neq 1,0}} h_k^2 + \sigma^2 .$$

A modification of Type B that cancels h_o, is [2], $e = h_1$ and

$$\hat{e}_i = \hat{a}_{i-1}\, (r_i - \hat{a}_i\, h_o) . \tag{A-3}$$

This is unbiased for error free conditions and

$$\sigma^2(\tau) = \sum_{k=-N}^{N} h_k^2 + \sigma^2.$$

Thus modified Type B has the lowest pattern jitter. The pattern jitter for Type A and modified Type B for $\sqrt{RC}$ R_x and T_x filters with $\alpha = 0.4$ and $\sigma^2 = 0$ was calculated and is given in Figure 15. The timing characteristic for modified Type B is shown in Figure 16. We note that this is not as linear or symmetric as the Type A case in Figure 14. Note that h_o must be estimated to realize the modified Type B algorithm in (6). As it is not linear and the pattern jitter for Types A and B (modified) are similar, we have chosen the Type A case in our BER simulations.

A-3 Timing Update

We show here that the timing update in equation (3) provides the averaging that results in $\langle e_k \rangle \longrightarrow e = h_1 - h_{-1}$. From (3) for $\tau_o = 0$,

$$\tau_N = \Delta \sum_{k=0}^{N-1} \hat{e}_i$$

where $\hat{e}_i$ is as in (A-1). For $\Delta = 1/N$,

$$\tau_N = \frac{1}{N} \sum_{k=0}^{N-1} \hat{e}_i$$

the sample mean of $\hat{e}_i$. Now as $\langle \hat{e}_i \rangle = e = h_1 - h_{-1}$, $\langle \hat{\tau}_N \rangle = e = h_1 - h_{-1}$, and thus

$$\Delta\tau_N = \tau_N - \langle \tau_N \rangle = \frac{1}{N} \sum_{k=0}^{N-1} (\hat{e}_k - e) .$$

Neglecting noise

$$\langle \Delta \tau_N^2 \rangle = \frac{1}{N} \left[\sigma^2(\tau) + \sum_{k=1}^{N-1} \Gamma_k(\tau) \right]$$

$$\Gamma_1(\tau) = h_o \, h_2 + h_o \, h_{-2} - R_2(\tau)$$

$$\Gamma_k = h_{k+1} \, h_{-(k-1)} + h_{k-1} \, h_{-(k+1)} - 2 \, h_k \, h_{-k}$$

$$\text{for } k > 1$$

$$R_2(\tau) = \sum_i h_i \, h_{i+2}$$

and $\sigma^2(\tau)$ is the pattern jitter in equation (A-2). Thus as N get large $\langle \Delta \tau_N^2 \rangle \longrightarrow 0$ and $\tau_N \longrightarrow e = h_1 - h_{-1}$, the required timing correction. This follows $h_i \sim c/i^3$ for i large and thus $\sum_{k=1}^{N-1} \Gamma_k(\tau)$ converges as $N \longrightarrow \infty$. We conclude that $\tau_N \longrightarrow e$ in the mean-square sense [6].

High Threshold (% of MAX Data Integrator)	# Out of Range (% of Total Sent)	# Bad (% of Total Sent)	# Missed (% of 180° Phase Shifts)
209	0	89	0
119	10	80	0
107	25	65	2
96	41	50	4
84	48	43	4
78	49	42	4
72	53	38	4
66	58	32	6
60	60	29	4
48	75	16	20
36	89	4	26

Table 1: Average Results over SNR Range of Interest (9 to 20 dB)

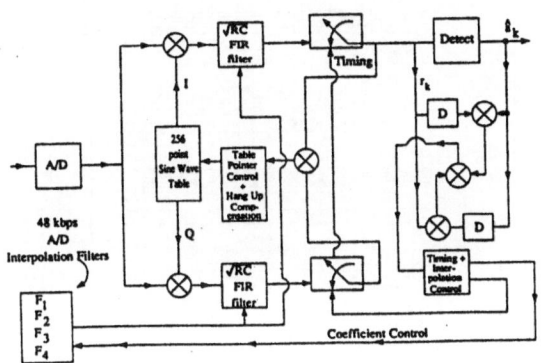

Figure 1: Receiver block diagram of coherent BPSK 6000 bps modem with polyphase coefficient interpolation timing recovery.

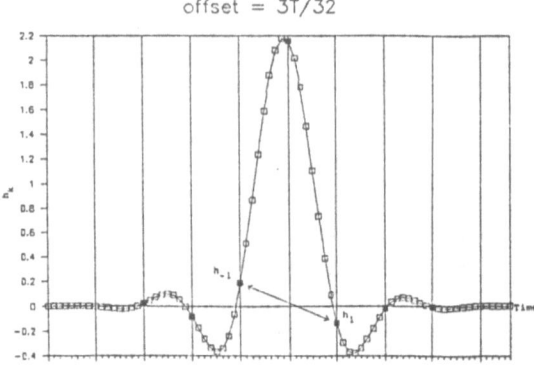

offset = 3T/32

Figure 2: Impulse response of 40% excess bandwidth raised cosine pulse, showing how timing error can be measured.

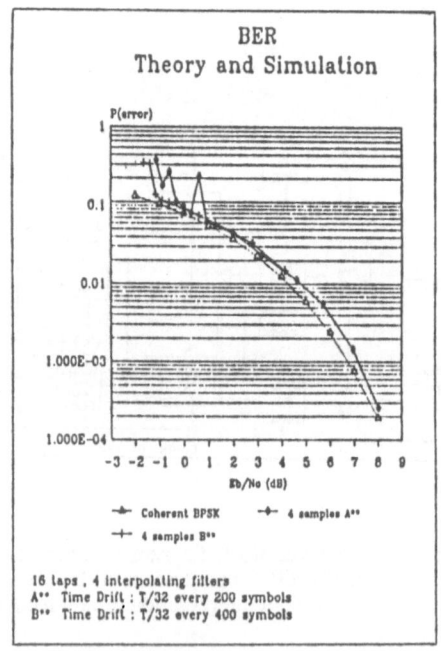

Figure 3: Simulation BER results for the 6000 bps bps coherent BPSK modem.

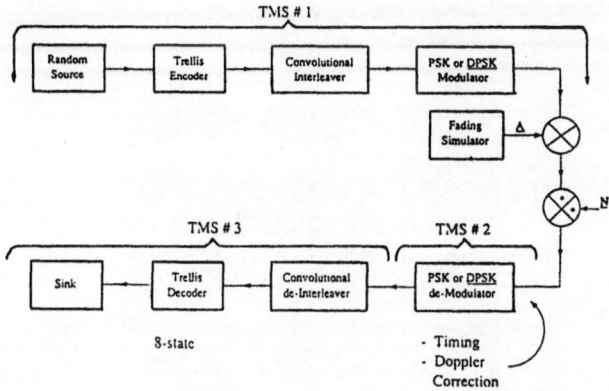

Figure 4: System block diagram MSAT modulation/ coding.

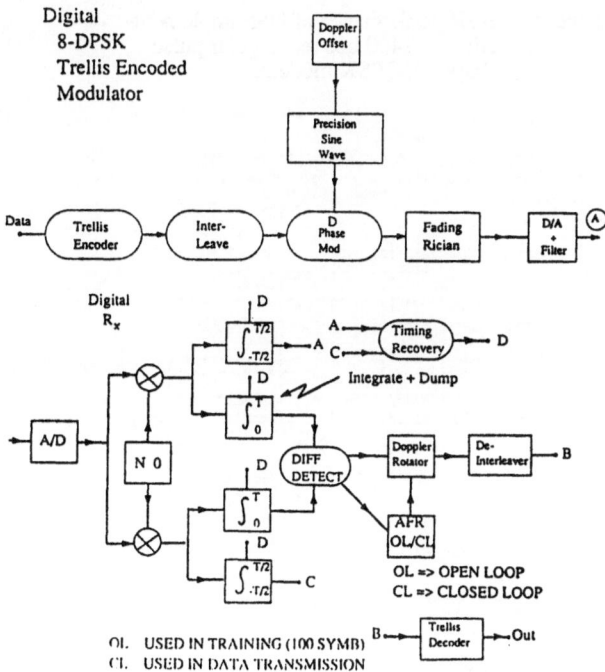

Figure 5: Transmitter and receiver block diagrams for 2400 bps rectangular pulse shaped DQPSK and 8-DPSK modems.

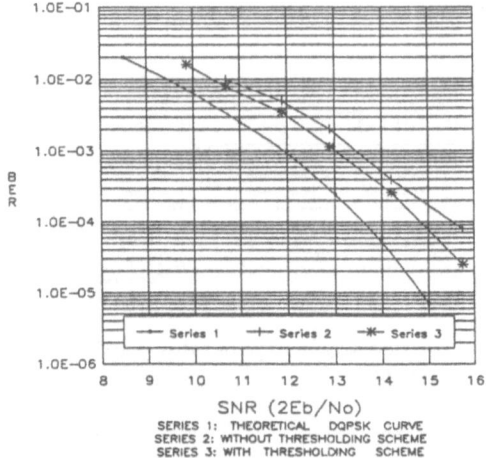

BIT ERROR RATE PERFORMANCE
UNCODED 2400 BPS. RECTANGULAR PULSE
SHAPED DQPSK.

SNR (2Eb/No)

SERIES 1: THEORETICAL DQPSK CURVE
SERIES 2: WITHOUT THRESHOLDING SCHEME
SERIES 3: WITH THRESHOLDING SCHEME

Figure 6: BER performance of DSP implementation of 2400 bps rectangular pulse
shaped DQPSK modem.

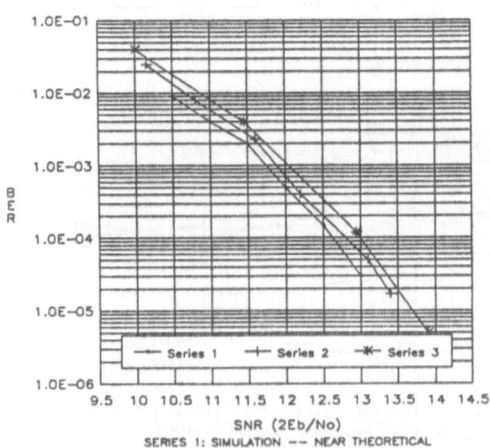

BIT ERROR RATE PERFORMANCE
RECTANGULAR PULSE SHAPED, INTERLEAVED
TCM 8-DPSK, 2400 BPS.

SNR (2Eb/No)

SERIES 1: SIMULATION -- NEAR THEORETICAL
SERIES 2: DSP WITH THRESHOLDING SCHEME
SERIES 3: DSP THRESHOLDING AND LOOP FILT

Figure 7: BER performance of 2400 bps TCM
8-DPSK modem.

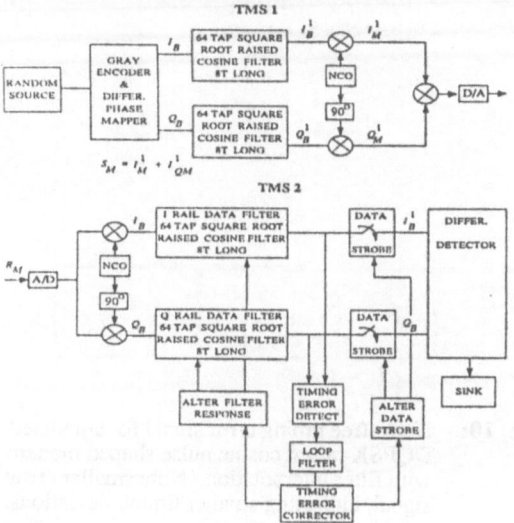

Figure 8: Transmitter and receiver block diagrams for raised cosine pulse shaped 4800 bps DQPSK modems.

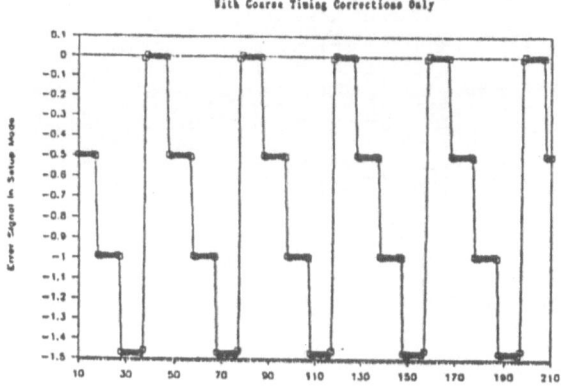

Figure 9: Noise free timing error signal for simulated DQPSK raised cosine pulse shaped modem, with no filter interpolation.

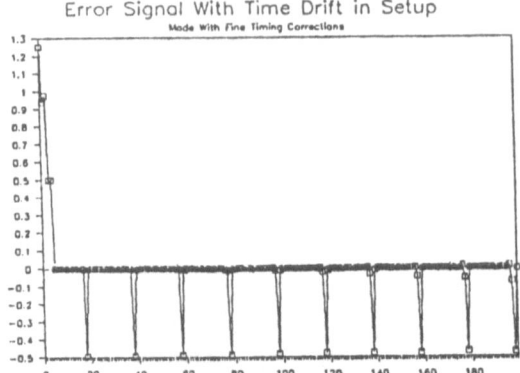

Figure 10: Noise free timing error signal for simulated DQPSK raised cosine pulse shaped modem with filter interpolation. Note smaller error signal, indicating smaller timing deviations.

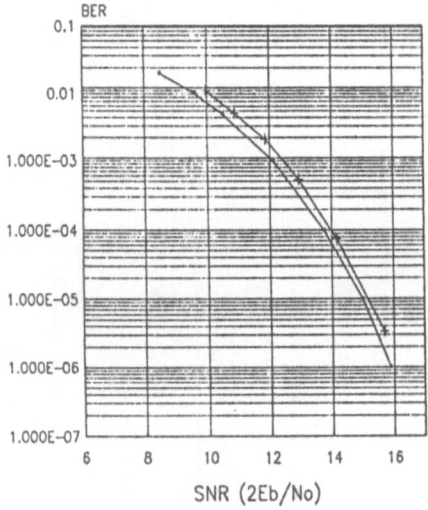

Figure 11: BER results for DSP implementation of 100% excess bandwidth pulse shaped 4800 bps DQPSK modem.

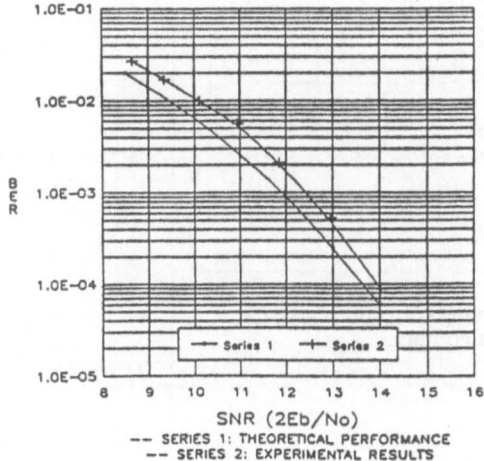

Figure 12: BER results for DSP implementation of 80% excess bandwidth pulse shaped 4800 bps DQPSK modem.

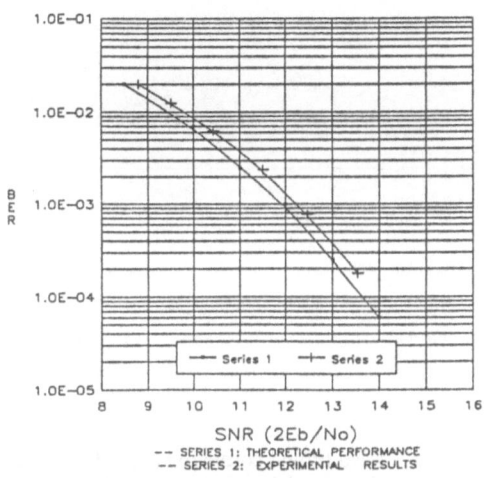

Figure 13: BER results for DSP implementation of 60% excess bandwidth pulse shaped 4800 bps DQPSK modem.

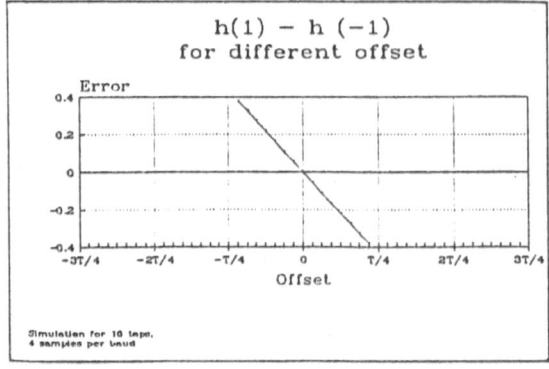

Figure 14: Ideal timing correction curve for M&M algorithm.

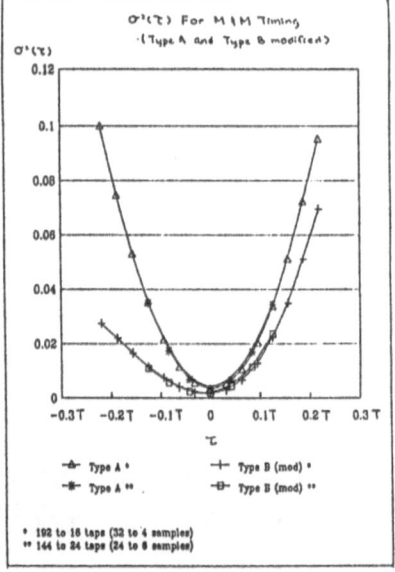

Figure 15: Pattern jitter for Type A and modified Type B M&M timing algorithms.

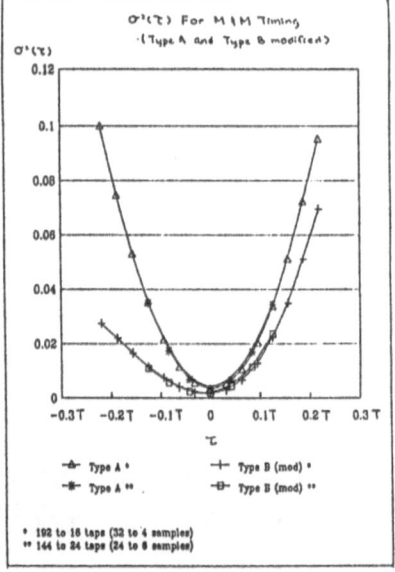

Figure 16: Ideal timing correction curve for modified Type B M&M timing algorithms.

Advanced modulation formats for satellite communications

Ezio Biglieri

Dipartimento di Elettronica, Politecnico di Torino
Corso Duca degli Abruzzi 24, I-10129 Torino (Italy)[1]

Abstract

We examine multidimensional modulation schemes in view of their application to digital satellite communications systems. In particular, we focus our attention on four-dimensional signals and describe several constellations with constant energy or carved from lattices.

1 Introduction

Motivation for the use of multidimensional signals in digital communication dates back to the early days of information theory, when Shannon recognized [21] that the performance of a signal constellation used to transmit digital information over the additive white Gaussian noise channel can be improved by increasing D, the dimensionality of the signal set used for transmission. In particular, as D grows to infinity, the performance tends to an upper limit which defines the capacity of the channel (see also [22]). Heuristically, as the number of dimensions grows we have more space to accommodate the signals, and hence the distance between signal points in the constellation increases. In turn, a greater distance between signal points means (at least for high enough signal-to-noise ratios) a smaller error probability.

We define the *power efficiency* of a signal constellation as

$$\Lambda = \frac{d_{\min}^2}{\mathcal{E}} \log_2 \mathcal{M},$$

where $d_{\min}^2$ is the minimum squared Euclidean distance among any pair of signals, $\mathcal{E}$ the average energy per symbol, and $\mathcal{M}$ is the number of signals available for transmission. The larger Λ, the better the error performance for the same (large) value of signal-to-noise ratio. A parameter that tells how efficiently the signal constellation makes use of the frequency is its *bandwidth efficiency*

$$R^* = \frac{\log_2 \mathcal{M}}{D}.$$

---

[1]Part of this work was sponsored by ESTEC.

It is measured in bits per dimension. The parameter R^* is meaningful for transmission over the additive white Gaussian noise channel, much less so in bandlimited channels. In fact, it does not depend on the actual signals used. The waveforms used over the bandlimited channel determine the bandwidth occupancy in Hz. For bandlimited channels, we define the bandwidth efficiency as

$$R = R^* \cdot \frac{D}{BT} = \frac{\log_2 \mathcal{M}}{BT} \tag{1}$$

measured in bits/sec/Hz (or bps/Hz). Here B is the bandwidth occupancy of the signals, and T^{-1} is the baud rate. This parameter depends on the specific waveform set chosen for transmission.

Some quantitative considerations on the efficiency of finite-dimensional signaling can be obtained by perusal of charts available in the literature. Fig. 1 shows sphere-packing upper bounds obtained by Slepian [22]. These (non-constructive) bounds show how certain performance parameters are related to signal dimensionality when coding is used. It can

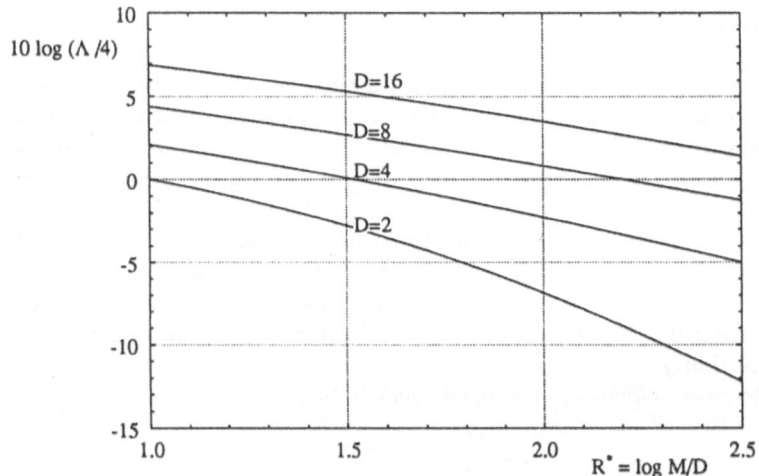

Figure 1: Sphere-packing upper bounds to the values of 10 log (Λ /4) achievable through coding and D-dimensional modulations.

be seen that for a given dimensionality D any increase of bandwidth efficiency entails a decrease of power efficiency, but also that for given R^* an increase in Λ can be achieved by increasing D. Fig. 1 shows the improvement in dB that is eventually offered by an increase of dimensionality (the reference value here is $\Lambda = 4$, corresponding to QPSK). For example, for $R^* = 2$, increasing the dimensionality from $D = 2$ to $D = 4$ a potential gain of about 5 dB can be achieved, provided of course that the "right" codes are found. Increasing D from 2 to 8 gives a potential gain of about 8 dB, and from 2 to 16, 10 dB. On the other hand, the increase of D also increases the complexity of the transmission system. In fact, the value $R^* = \log_2 M/D = 2$ implies a constellation of $M = 2^{2D}$ signals, i.e., 16 signals in 2 dimensions, 256 signals in 4 dimensions, and 65536 signals in 8 dimensions.

Another conclusion that can be inferred from Fig. 1 is that the principle of diminishing returns holds for this problem: the maximum benefit when dimensions are added is achieved from 2 to 4 dimensions, while any further increase of D yields reduced benefits. This seems to show that good results can be expected by using four-dimensional signals in conjunction with coding. This observation is further reinforced by looking at Fig. 2, adapted from [9]. This figure plots the performance of *actual designs* of TCM schemes based on D-dimensional signals, $D = 2, 4, 6, 8$, versus their complexity, defined (following Forney [14]) as the number of operations — additions, subtractions, comparisons — per symbol period in the decoder. The number of information bits per symbol is taken equal to 6 in the chart. It is interesting to observe that, for coding gains above 4, 4-dimensional TCM schemes offer the best performance. In [9], this is explained by observing that by increasing the number of dimensions we do increase the asymptotic coding gain, but also the error coefficient. We quote from [9]: "Beyond some dimension, the increase of the error coefficient becomes dominant, and the net coding gain at practical P_e values becomes smaller as the dimension increases."

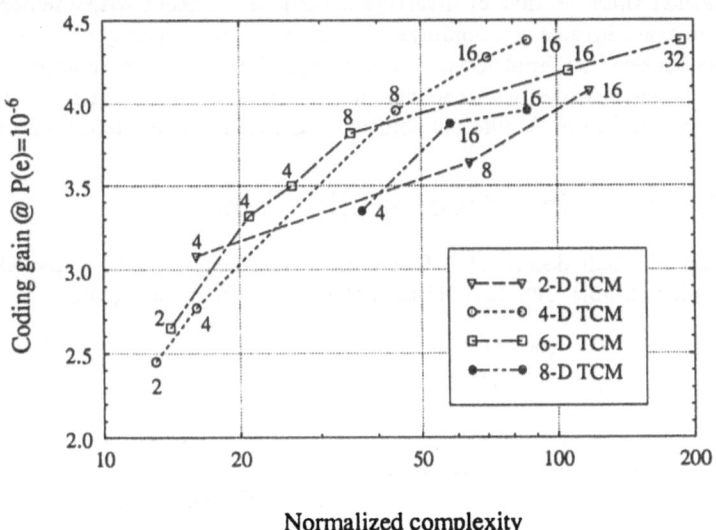

Normalized complexity

Figure 2: Comparison of 2-dimensional and multidimensional TCM schemes at the error rate of 10^{-6} for 6 information bits per symbol. Each point in this chart represents an actual design. The label of each point is the number of states of the design.

Until recently, signal sets with dimensionality larger than 2 were not considered for practical applications. Four-dimensional signals were considered, among others, by Welti and Lee [24], Zetterberg and Brändström [27], Wilson *et al.* [25, 26], and Biglieri and Elia

[3, 4]. Gersho and Lawrence [17] consider four- and eight-dimensional signal sets with two information bits per dimension. Their designs show a 1.2 dB to 2.4 dB gain in noise margin over conventional (two-dimensional) quadrature amplitude modulation.

1.1 Impairments in satellite channels

Besides the effect of additive noise and of intersymbol interference, one should take into account the fact that often power amplifiers in satellite systems are operated at or near saturation for better efficiency. Signal constellations whose envelope is not constant suffer from the channel nonlinearities, and hence nonlinear distortions may prevent the use of certain signal sets if suitable countermeasures are not taken.

A simple solution is memoryless data predistortion. It consists of pre-warping the signal constellation so that after amplifier distortion it will be restored to its original shape [19]. Now, in the presence of channel memory this countermeasure may not be effective enough. One may resort to channel equalization based on nonlinear equalizers with memory [1], which on the other hand could result into noise enhancement. A better proposition seems to be offered by data predistortion with memory, which operates on the transmitter side and hence on a noiseless signal. Data predistortion lends itself to digital implementation and can be easily made adaptive.

A data predistortion technique with memory was proposed in [2]. The predistortion problem was regarded there as that of inverting a nonlinear system with memory (the channel). Since the exact inverse of a nonlinear system with memory may not exist, in [2] a predistorter was chosen that implements the pth-order inverse of the channel. This is the system which, when cascaded with the channel, removes all the nonlinearities whose order is less than or equal to p. A different solution has recently been described in [18].

2 Q^2PSK and related constellations

Recently, Saha and Birdsall described a four-dimensional signal basis, called Q^2PSK, which is easily implementable [20]. It consists of the four orthogonal signals

$$p(t) \cos 2\pi f_c t \qquad q(t) \cos 2\pi f_c t$$
$$p(t) \sin 2\pi f_c t \qquad q(t) \sin 2\pi f_c t, \tag{2}$$

where $0 \leq t < T$, $p(t)$ and $q(t)$ are the orthogonal sinusoids $p(t) = \cos(\pi t/T)$, $q(t) = \sin(\pi t/T)$, and f_c denotes the carrier frequency, with $f_c \gg 1/T$. The four-dimensional vector $\mathbf{a} = (a_1, a_2, a_3, a_4)$ can be transmitted with this signal basis in the form

$$\begin{aligned} s(t, \mathbf{a}) &= a_1 p(t) \cos 2\pi f_c t + a_2 q(t) \cos 2\pi f_c t \\ &+ a_3 p(t) \sin 2\pi f_c t + a_4 q(t) \sin 2\pi f_c t. \end{aligned} \tag{3}$$

If a_i, $i = 1, \cdots, 4$, are uncorrelated identically distributed zero-mean random variables, the power spectral density of the digital signal obtained by transmitting a sequence of $\mathbf{a}$-vectors is proportional to

$$\operatorname{sinc}^2(fT + 1/2) + \operatorname{sinc}^2(fT - 1/2),$$

where $\operatorname{sinc}(x) = \sin(\pi x)/\pi x$.

Now, compare the Q^2PSK signal basis with the four-dimensional basis obtained by pairing two quadrature sinusoids with duration $T/2$ (QPSK) so that

$$s(t, \mathbf{a}) = a_1 c(t) + a_2 s(t) + a_3 c(t - T/2) + a_4 s(t - T/2),$$

where $c(t) = \cos(2\pi f_c t)$ and $s(t) = \sin(2\pi f_c t)$, $0 \leq t < T/2$. The power spectral density obtained in this case is proportional to $\mathrm{sinc}^2(fT/2)$. It can be seen that Q^2PSK uses the spectrum more efficiently than QPSK (at least for large enough bandwidths), but it should be kept in mind that this occurs at the expense of a *non-constant* envelope.

2.1 Optimized Q^2PSK

Q^2PSK can be modified for digital transmission over severely bandlimited channels. A new version of Q^2PSK [23] is based on the derivation of a set of four-dimensional basis signals that retain the form (2), are orthogonal, and whose bandwidth occupancy is minimal. The assumption of limited bandwidth, and consequently of time-unlimited signals, forces a definition of orthogonality that generalizes its classical definition.

Assume that a (multidimensional) waveform is transmitted every T seconds, and that the receiver operates by sampling the observed signal at rate $1/T$. We design a set of basis signals whose duration may exceed T, and satisfy the following requirements:

1. When passed through their matched filter, they produce an output sample sequence that has no intersymbol interference (ISI).

2. When passed through the filter matched to another basis signal, they produce an output sample sequence that has no interchannel interference (ICI).

3. Their bandwidth occupancy is minimum.

The transmitted signal retains the form (3).

At the receiver front-end (see Fig. 3), multiplication by $\cos 2\pi f_c t$ and $\sin 2\pi f_c t$, followed by lowpass filtering, separates the two pairs of dimensions. Further, it is necessary to separate a_{1n} from a_{2n}, and a_{3n} from a_{4n}. This is done by using a pair of filters matched to $p(t)$ and $q(t)$, respectively. In the absence of noise (and of ISI) the output of the first filter should be a_{1n}, and that of the second filter a_{2n}. Moreover, we want the two noise components at the output of the matched filters to be independent under the usual assumption of a white Gaussian channel noise.

2.1.1 Continuous signal model

Consider again the orthogonality conditions listed before. Let $y_{pp}(t)$ be the response of the filter matched to $p(t)$ when the input signal is $p(t)$, $y_{qq}(t)$ the response of the filter matched to $q(t)$ when the input signal is $q(t)$, and $y_{pq}(t)$ the response of the filter matched to $p(t)$ when the input signal is $q(t)$. If $\mathcal{F}$ denotes Fourier transform, and $\mathcal{F}[p(t)] = P(f)$, $\mathcal{F}[q(t)] = Q(f)$, then we have

$$\mathcal{F}[y_{pp}(t)] = |P(f)|^2, \tag{4}$$

$$\mathcal{F}[y_{qq}(t)] = |Q(f)|^2, \tag{5}$$

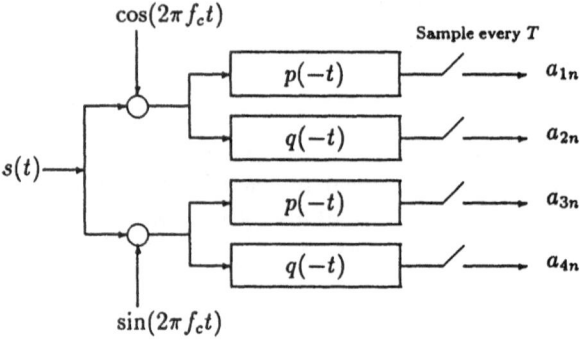

Figure 3: Receiver structure

and

$$\mathcal{F}[y_{pq}(t)] = P(f)Q^*(f). \tag{6}$$

Now, denote by $\Delta_T(t)$ the sampling function, i.e., a periodic train of pulses spaced by T:

$$\Delta_T(t) = \sum_{n=-\infty}^{\infty} \delta(t - nT).$$

The conditions for absence of ISI and ICI for the output of the four matched filters, as sampled every T, can be expressed in the form

$$y_{pp}(t) \cdot \Delta_T(t) = \delta(t) \qquad \text{(no ISI)} \tag{7}$$

$$y_{qq}(t) \cdot \Delta_T(t) = \delta(t) \qquad \text{(no ISI)} \tag{8}$$

$$y_{pq}(t) \cdot \Delta_T(t) = 0 \qquad \text{(no ICI)}. \tag{9}$$

By taking the Fourier transforms of both sides of (7)-(9), recalling (4)-(5) we get the conditions

$$\sum_{n=-\infty}^{\infty} |P(f - n/T)|^2 = \text{a constant} \qquad \text{(no ISI)}, \tag{10}$$

$$\sum_{n=-\infty}^{\infty} |Q(f - n/T)|^2 = \text{a constant} \qquad \text{(no ISI)}, \tag{11}$$

$$\sum_{n=-\infty}^{\infty} P(f - n/T)Q^*(f - n/T) = 0 \qquad \text{(no ICI)}. \tag{12}$$

Eqs. (10) and (11) simply express the first Nyquist condition for the spectra $|P(f)|^2$ and $|Q(f)|^2$. The new fact here is the condition (12) of no ICI, which further constrains the actual shapes of $P(f)$ and $Q(f)$. Fig. 4 shows a minimum-bandwidth pair of spectra $P(f)$, $Q(f)$.

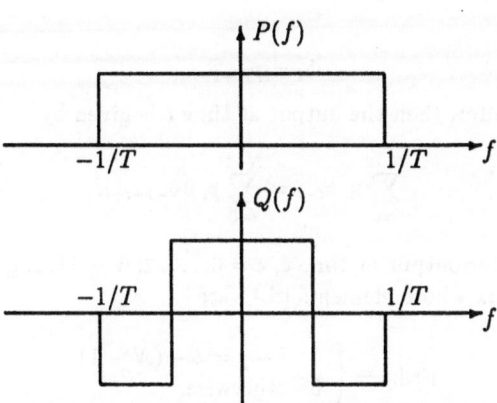

Figure 4: Fourier transforms of the signals $p(t)$ and $q(t)$ that satisfy the orthogonality conditions.

Some considerations. From Fig. 4 we see that the minimum bandwidth occupancy of the four-dimensional ISI-free, ICI-free signal (3) is $2/T$ Hz, which corresponds to the maximum transmission rate of 2 bps/Hz. This is the same as for quaternary PSK with pulses designed for minimum bandwidth. Consequently, for transmission over the ideal (i.e., distortionless) additive white Gaussian noise channel there is no improvement in throughput by using uncoded Q^2PSK rather than QPSK. In fact, the bandwidth occupancy is the same, as is the error probability. To get a benefit from a four-dimensional basis on this ideal channel a coding scheme should be used.

2.1.2 Discrete signal model

To generate a discrete-time model for the signal basis developed above we use the technique developed in [8]. We assume that the sequence of transmitted waveforms be generated by sending the information symbols, one every T seconds, into an N-tap finite-impulse-response transmission filter. This filter is clocked at rate I/T corresponding to an interpolation (or oversampling) factor I. Thus, we represent the waveforms $p(t)$ and $q(t)$ in the form of two column vectors

$$\mathbf{p} = [p_0, p_1, \cdots, p_{N-1}]^T$$

and

$$\mathbf{q} = [q_0, q_1, \cdots, q_{N-1}]^T$$

(where the superscript ^{T} denotes transpose) which extend over $M = \lceil N/I \rceil$ symbol intervals.

Matched filtering. Consider the filter matched to $\mathbf{p}$. It has impulse response

$$h_0, \ h_1, \ \cdots, \ h_{N-1}$$

with

$$h_i = p_{N-1-i}.$$

If **p** is sent into this filter, then the output at time ℓ is given by

$$\sum_{i=0}^{N-1} p_i \, h_{\ell-i} = \sum_{i=0}^{N-1} p_i \, p_{N-1-\ell+i}.$$

Thus, the matched filter output at time ℓ, $\ell = 0, \ldots, 2(N-1)$, is given by $\mathbf{p}^T \mathbf{S}_\ell \mathbf{p}$, where $\mathbf{S}_\ell$ is the $N \times N$ matrix whose elements $[\mathbf{S}_\ell]_{ij}$ are

$$[\mathbf{S}_\ell]_{ij} = \begin{cases} 1 & i - j = \ell - (N-1) \\ 0 & \text{otherwise.} \end{cases}$$

In particular, $\mathbf{S}_{N-1}$ is the $N \times N$ identity matrix, and

$$\mathbf{S}_\ell^T = \mathbf{S}_{2(N-1)-\ell}. \tag{13}$$

Similarly, the response to **q** of the filter matched to the same signal is $\mathbf{q}^T \mathbf{S}_\ell \mathbf{q}$. The response to **p** of the filter matched to **q** is $\mathbf{p}^T \mathbf{S}_\ell \mathbf{q}$, and finally the response to **q** of the filter matched to **p** is $\mathbf{q}^T \mathbf{S}_\ell \mathbf{p}$.

2.1.3 Signal optimization

The optimization criterion that we choose is the minimization of bandwidth occupancy. The signal power concentrated in the frequency interval $|f| \leq B$ is given by [8]

$$\begin{aligned} J &= \int_{-B}^{B} \left[|P(f)|^2 + |Q(f)|^2 \right] df \\ &= \mathbf{p}^T \mathbf{R} \mathbf{p} + \mathbf{q}^T \mathbf{R} \mathbf{q}, \end{aligned} \tag{14}$$

where $\mathbf{R}$ is an $N \times N$ symmetric Toeplitz matrix with elements

$$r_{ik} = \frac{\sin(2\pi B(i-k)T/I)}{\pi(i-k)T/I}.$$

No intersymbol interference. Information symbols are transmitted every I sampling instants. Consequently, for the output of the two matched filters at time $kI + N - 1$ ($k \in \{-M, \cdots, M\}$, and $N - 1$ is the delay introduced by the cascade of the transmitter and receiver filters), to depend only on the kth information symbol, i.e., for no intersymbol interference at time $kI + N - 1$, we must have

$$\mathbf{p}^T \mathbf{S}_{kI+N-1} \mathbf{p} = \delta(k), \tag{15}$$

and

$$\mathbf{q}^T \mathbf{S}_{kI+N-1} \mathbf{q} = \delta(k). \tag{16}$$

No interchannel interference. The second constraint that we impose is the absence of interchannel interference (ICI) at the sampling times $kI + N - 1$, $(k \in \{-M, \cdots, M\})$, when p is passed through the filter matched to q and when q is passed through the filter matched to p:

$$\mathbf{p}^T \mathbf{S}_{kI+N-1} \mathbf{q} = 0, \tag{17}$$

and

$$\mathbf{q}^T \mathbf{S}_{kI+N-1} \mathbf{p} = 0. \tag{18}$$

By taking the transpose of both sides of (18), and recalling that the transpose of $\mathbf{S}_\ell$ is $\mathbf{S}_{2(N-1)-\ell}$, we see that the condition (18) is automatically satisfied if (17) is satisfied.

Independent noise samples. Let n_h be the sample at instant h of the additive white Gaussian noise process $\{n_h\}_{h=-\infty}^{\infty}$.

When the noise process enters the filter matched to p, the h-th sample of the output process is

$$m_h(\mathbf{p}) = \sum_{i=0}^{N-1} p_i n_{h-N+1+i} = \mathbf{p}^T \tilde{\mathbf{n}}$$

where $\tilde{n}_i = n_{h-N+1+i}$. In the same way, the output of the filter matched to q is $m_h(\mathbf{q}) = \mathbf{q}^T \tilde{\mathbf{n}}$.

Since the random variables $m_h(\mathbf{p})$ and $m_h(\mathbf{q})$ are Gaussian, to prove their independence it is sufficient to show that they are uncorrelated, which can be done in the following way:

$$E\{m_h(\mathbf{p})m_h(\mathbf{q})\} = \mathbf{p}^T E\{\tilde{\mathbf{n}}\tilde{\mathbf{n}}^T\} \mathbf{q} = \sigma^2 \mathbf{p}^T \mathbf{q} = 0.$$

This result holds whenever the samples of the noise process are independent and identically distributed, and the two vectors p and q are orthogonal. That p and q are actually orthogonal is shown by letting $k = 0$ in (17).

The solution. Employing elementary variational calculus techniques, we must maximize

$$\frac{\mathbf{p}^T \mathbf{R} \mathbf{p} + \mathbf{q}^T \mathbf{R} \mathbf{q}}{\mathbf{p}^T \mathbf{p} + \mathbf{q}^T \mathbf{q}}$$

under the three constraints (15), (16), and (17). This can be done by using the projected gradient method, as suggested in [8].

2.1.4 Results and comparisons

It can be seen [23] that optimized Q^2PSK achieves the same bandwidth occupancy as optimized QPSK [8] with a lower complexity. This complexity gain is more significant when the bandwidth is smaller. Further, consider transmission over a nonlinear bandlimited channel typical of satellite transmission. The nonlinearity is generated by a typical traveling-wave tube amplifier working at or near saturation. In [23], a comparison is made between optimized QPSK and Q^2PSK. It is shown that for a large bandwidth occupancy optimized Q^2PSK is just marginally more power-efficient than QPSK, while it outperforms QPSK by a couple of dB for narrow-band transmission. Thus, without coding the use of optimized four-dimensional signals basis may prove advantageous over satellite channels. Use of coded modulation may further enhance its advantages.

Code	$\mathcal{M}$	$d^2_{min}/\mathcal{E}$	$d^2_{next}/\mathcal{E}$	Λ	R^*	power gain (dB)	bandwidth expansion
Simplex	5	2.5	∞	5.8	0.58	1.6	1.7
Biorthogonal	8	2	4	6	0.75	1.8	1.33
Q²PSK	16	1	2	4	1	0	1
24-cell	24	1	2	4.59	1.15	0.6	0.87
48-cell	48	0.59	1	3.27	1.4	-0.9	0.71
120-cell	120	0.38	1	2.64	1.73	-1.8	0.58

Table 1: Some four-dimensional signal constellations and their parameters. Power gain and bandwidth expansion are relative to Q²PSK, i.e, the uncoded constellation.

3 Constant-energy 4D signal sets

In this section we examine some examples of signal sets wrapped on the surface of a 4-dimensional sphere. Clearly, the equal-energy constraint would result in signal sets with lower energy efficiencies at the expense of a low peak-energy to average-energy ratio. Our comparison is based of the values of the parameter pair (Λ, R^*) associated with each one of the constellations. The list of signal sets shown below is obviously not exhaustive, but will provide a feeling of the variety of available tradeoffs between power and bandwidth efficiencies.

Table 3, elaborated from the results of [6, 27], shows the tradeoffs between power and bandwidth efficiency that can be obtained by selecting a signal constellation in the four-dimensional space. The codes considered there have constant energy, i.e., the signal points lie on the surface of the four-dimensional sphere with radius $\sqrt{\mathcal{E}}$. For each constellation, we show the values of d^2_{min}, the squared minimum Euclidean distance, and of d^2_{next}, the squared second largest Euclidean distance. The latter value is useful to assess the constellation performance at intermediate values of SNR. The performance of these constellations is plotted in the (Λ, R^*)-chart shown in Fig. 5.

4 Signal sets based on lattices

An important (but non-constructive) result by deBuda [13] states that over the ideal additive white Gaussian noise channel *lattice constellations* can essentially approach capacity. In fact, given any transmission rate below capacity and any small error probability ϵ, a D-dimensional lattice constellation exists with an error probability smaller than ϵ. This constellation consists of all the points of some lattice lying within a D-dimensional sphere. Here we consider 4-dimensional signal constellations based on *lattices*. (A thorough treatment of this topic can be found in the book by Conway and Sloane [10]).

In general, an N-dimensional lattice $\mathcal{L}$ is defined as an infinite set of points, or N-vectors, closed under ordinary addition and multiplication by integers. If d_{min} is the minimum distance between any two points in the lattice, the *kissing number* τ is the number of adjacent lattice points located at distance d_{min}, i.e., the number of nearest neighbors of any lattice point.

Name	$\mathcal{L}$	N	$\gamma_c(\mathcal{L})$ (dB)
Integer lattice	$\mathbf{Z}$	1	0.00
Hexagonal lattice	A_2	2	0.62
Schläfli	D_4	4	1.51
Gosset	E_8	8	3.01
Barnes-Wall	Λ_{16}	16	4.52
Leech	Λ_{24}	24	6.02

Table 2: Coding gains of lattices.

4.1 Interesting four-dimensional lattices

The lattice $\mathbf{Z}^4$. The set $\mathbf{Z}^4$ of all 4-tuples with integer coordinates is called the *cubic lattice*, or *integer lattice*. The minimum distance is $d_{\min} = 1$, and the kissing number is $\tau = 8$.

The lattice A_4. This is the set of all vectors with 5 integer coordinates whose sum is zero. This lattice may be viewed as the intersection of $\mathbf{Z}^5$ and a hyperplane cutting the origin. The minimum distance is $d_{\min} = \sqrt{2}$, and the kissing number is $\tau = 20$.

The lattice D_4. This is the set of all 4-dimensional points whose integer coordinates have an even sum. It may be viewed as a punctured version of $\mathbf{Z}^4$, in which the points are colored alternately red and white with a checkerboard coloring, and the white points (those with odd sums) are removed. We have $d_{\min} = \sqrt{2}$, and $\tau = 24$.

D_4 represents the densest lattice packing in $\mathbf{R}^4$. This means that if unit-radius, 4-dimensional spheres with centers in the lattice points are used to pack $\mathbf{R}^4$, then D_4 is the lattice with the largest number of spheres per unit volume.

4.2 The coding gain of a lattice

The *coding gain* $\gamma_c(\mathcal{L})$ of the N-dimensional lattice $\mathcal{L}$ is defined as

$$\gamma_c(\mathcal{L}) = \frac{d_{\min}^2(\mathcal{L})}{V(\mathcal{L})^{2/N}}, \tag{19}$$

where $V(\mathcal{L})$ is the *fundamental lattice volume*, i.e., the reciprocal of the number of lattice points per unit volume (for example, $V(\mathbf{Z}^N) = 1$). The main properties of $\gamma_c(\mathcal{L})$ are listed in [14, pp. 1128-1129]. Table 2 lists the coding gains of some among the most popular lattices.

4.3 Carving a signal constellation out of a lattice

In this section we shall study how a finite signal constellation C can be obtained from an infinite 4-dimensional lattice $\mathcal{L}$, and some of the properties of the resulting constellation.

We shall denote with $C(\mathcal{L}, \mathcal{R})$ a constellation obtained from $\mathcal{L}$ (or from its translate $\mathcal{L}+\mathbf{a}$) by retaining only the points that fall in the region $\mathcal{R}$. The resulting constellation has

$$\mathcal{M} \approx \frac{V(\mathcal{R})}{V(\mathcal{L})}$$

points, provided that $V(\mathcal{R}) \gg V(\mathcal{L})$, i.e., that $|C|$ is large enough.

In order to quote a result that expresses the quality of the constellation C, define its "figure of merit" as

$$\text{CFM}(C) = 2\frac{d_{\min}^2}{\mathcal{E}}.$$

Then, define the *shape gain* $\gamma_s(\mathcal{R})$ of the region $\mathcal{R}$ as the ratio between the *normalized second moment* of any 4-dimensional cube (which is equal to $1/12$) and the normalized second moment of $\mathcal{R}$:

$$\gamma_s(\mathcal{R}) = \frac{1/12}{G(\mathcal{R})}, \tag{20}$$

where

$$G(\mathcal{R}) = \frac{\int_{\mathcal{R}} \|\mathbf{r}\|^2 \, d\mathbf{r}}{NV(\mathcal{R})^{3/2}}, \tag{21}$$

and V is the volume of $\mathcal{R}$. The main properties of $\gamma(\mathcal{R})$ are listed in [15, p. 884].

We can now quote the following result [15]:

> The figure of merit of the constellation $C(\mathcal{L}, \mathcal{R})$ having normalized bit rate β is given by
>
> $$\text{CFM}(C) \approx \frac{6}{2^\beta} \cdot \gamma_c(\mathcal{L}) \cdot \gamma_s(\mathcal{R}), \tag{22}$$
>
> where $6/2^\beta$ is the figure of merit of the one-dimensional PAM constellation (chosen as the baseline), $\gamma_c(\mathcal{L})$ is the coding gain of the lattice $\mathcal{L}$ (see 19), and $\gamma_s(\mathcal{R})$ is the shaping gain of the region $\mathcal{R}$. The approximation holds for large constellations.

This result shows that, at least for large constellations, the gain from shaping by the region $\mathcal{R}$ is almost completely decoupled from the coding gain due to $\mathcal{L}$ — or, more generally, the gain due to the use of a code. Thus, for a good design it makes sense to optimize separately $\gamma_c(\mathcal{L})$ (i.e., the choice of the lattice) and $\gamma_s(\mathcal{R})$ (i.e., the choice of the region).

4.3.1 Spherical constellations

The maximum shape gain achieved by a 4-dimensional region $\mathcal{R}$ is that of a sphere Σ. If R is its radius, it has

$$V(\Sigma) = \frac{\pi^2 R^4}{2}$$

and

$$\int_\Sigma \|\mathbf{r}\|^2 \, d\mathbf{r} = \frac{2}{3} R^2 V(\Sigma),$$

so that

$$\gamma_s(\Sigma) = \frac{\pi}{2\sqrt{2}} \approx 1.11.$$

4.3.2 Voronoi constellations

Voronoi constellations were introduced in [11] and further studied in [16]. Given a lattice $\mathcal{L}$ and its sublattice $\mathcal{L}'$, form the partition $\mathcal{L}/\mathcal{L}'$. The *Voronoi region* $\mathcal{R}_V(\mathcal{L}')$ is the set of points that are at least as close to the origin 0 as to any other point in $\mathcal{L}'$. A Voronoi constellation is obtained by choosing the points of a translate $\mathcal{L} + a$ that fall in the region $\mathcal{R} = \mathcal{R}_V(\mathcal{L}')$. Formally, we have

$$C \quad \text{consists of all vectors } x - a \quad \text{for } x \in \mathcal{L} \cap (a + \mathcal{R}_V(\mathcal{L}')). \tag{23}$$

Another way of characterizing a Voronoi constellation is the following. Every translate $\mathcal{L} + a$ of $\mathcal{L}$ is a union of $|\mathcal{L}/\mathcal{L}'|$ cosets of $\mathcal{L}'$. In every coset, choose the element with the minimum norm. The set of these minimum-norm points is the Voronoi constellation (see later about the problem of choosing among elements with the same norm).

A Voronoi constellation based on the lattice $\mathcal{L}$ and its partition $\mathcal{L}/\mathcal{L}'$ has $|\mathcal{L}/\mathcal{L}'|$ points, with a chosen so as to minimize the average constellation energy.

The shape gain for the Voronoi regions of the "Schläfli" lattice D_4 is $\gamma_s = 0.37$ dB, while $\gamma_\Sigma = 0.46$ dB.

How to choose a Voronoi constellation. Given $\mathcal{L}$ and $\mathcal{L}'$, to specify a Voronoi constellation based on the partition $\mathcal{L}/\mathcal{L}'$ it is necessary to specify the translation N-tuple a, and to resolve any ties when a coset of $\mathcal{L}'$ in $\mathcal{L}$ has more than one minimum-norm element. Three methods are available to do this, viz.,

1. The "Conway-Sloane method," described in [11].

2. The "Maximally biased method (Forney's method 1)", described in [16, p. 942-943].

3. "Forney's method 2", described in [16, p. 942-943].

Labeling Voronoi constellations. As mentioned before, in practice it is necessary to have a method of mapping data words to constellation points and vice versa. Surprisingly enough, the complexity of this operation is comparable to the complexity of the demodulation. Two such methods are known. One is due to Conway and Sloane [11], while a generalization is described in [16].

4.3.3 Nonequiprobable constellations

A different approach to the problem of carving a constellation out of a lattice has been recently taken by Calderbank and Ozarow in [7]. In this paper a signal constellation is partitioned into T subconstellations of equal size by scaling a basic region $\mathcal{R}$. Signal points in the same subconstellation are used with the same probability, and a shaping code selects each subconstellation with a certain frequency. For comparable shape gain and constellation expansion ratio the peak-to-average power ratio of these schemes is superior to Voronoi constellations. Moreover, table look-up is all that is required to address points, while in Voronoi constellations the encoding complexity is governed by the complexity of decoding the lattice.

4.4 Decoding procedures for 4-dimensional lattices

Let $\mathbf{r} = (r_1, r_2, r_3, r_4)$ denote the received vector. For maximum-likelihood detection we must find the lattice point with the minimum Euclidean distance from $\mathbf{r}$. Simple procedures have been obtained in [10] (see also [12, pp. 442 ff.]). Notice that the methods described presume an infinite lattice, so that it may happen that the decoded lattice point does not belong to the constellation. If this happens, countermeasures must be taken which depend on the specific constellation shape.

4.5 Asymptotics of lattice signal designs

In four dimensions, it is known that the densest packing of spheres is provided by the lattice D_4. This suggests that D_4 will produce optimal signal constellations for high signal-to-noise ratio additive white Gaussian noise channels.

For spherical constellations, with large $\mathcal{M}$ the ratio of squared minimum distance to average energy approaches [24]

$$\frac{d_{min}^2}{\mathcal{E}} = \frac{3\pi}{2\sqrt{\mathcal{M}}}.$$

Thus, the power efficiency of a spherical constellation carved from D_4 is

$$\Lambda = \frac{3\pi}{2} \frac{\log_2 \mathcal{M}}{\sqrt{\mathcal{M}}}.$$

For example, with $\mathcal{M} = 64$ we have

$$\Lambda = \frac{3\pi}{2} \frac{6}{8} \approx 3.53$$

with

$$R^* = \frac{6}{4} = 1.5.$$

Thus, this modulation scheme is only 0.3 dB less power-efficient than QPSK, but is 50% spectrally more efficient. At $\mathcal{M} = 1024$, we have $\Lambda = 1.47$ and $R^* = 2.5$, that is, a scheme which is 4.3 dB worse than QPSK in terms of power efficiency, but 2.5 times spectrally more efficient.

Since the packing density of Z^4 is half of that of D_4, while that of A_4 is 89% of that of D_4, for large $\mathcal{M}$ the power efficiencies of these three 4-dimensional lattices are in the same ratio.

5 Design of spherical constellations

In practical systems we are interested in signal constellations whose size is a power of two, so that an integer number of bits are carried by each symbol. To achieve this goal with spherical constellations, we enumerate the spherical shells and the number of points in each shell for Z^4, D_4, and A_4. It may prove useful, in terms of energy efficiency, to move the origin within the lattice, i.e., to consider the lattice translate $\mathcal{L} + \mathbf{a}$.

Shell #	Energy of shell	Signals on shell	Signals within shell
1	0.	1	1
2	1.	8	9
3	2.	24	33
4	3.	32	65
5	4.	24	89
6	5.	48	137

Table 3: Shell count for $\mathbf{Z}^4$.

We now describe some results, derived in [5], useful for enumerating the shells and the signals within each shell. This technique is based on the derivation of the transfer function of a trellis describing the lattice. It has the form

$$T(\mathcal{L}; D) = a_\alpha D^\alpha + a_\beta D^\beta + \cdots,$$

and tells us that there are a_α lattice points with energy α, a_β lattice points with energy β, etc.

Define the function

$$g_s(D) = D^{s^2} + D^{(1-s)^2} + D^{(1+s)^2} + \cdots = \sum_{n=-\infty}^{\infty} D^{(s+n)^2}.$$

Then the enumerator of $\mathbf{Z}^4 + (a, b, c, d)$ is

$$T(\mathbf{Z}^4 + (a, b, c, d); D) = g_a(D) \times g_b(D) \times g_c(D) \times g_d(D), \qquad (24)$$

while the enumerator of $D_4 + (a, b, c, d)$ is

$$
\begin{aligned}
& g_a(D)g_b(D)g_c(D)g_d(D) + g_a(D)g_b(D)g_{1-c}(D)g_{1-d}(D) + \\
+ \; & g_a(D)g_{1-b}(D)g_c(D)g_{1-d}(D) + g_a(D)g_{1-b}(D)g_{1-c}(D)g_d(D) + \\
+ \; & g_{1-a}(D)g_b(D)g_c(D)g_{1-d}(D) + g_{1-a}(D)g_b(D)g_{1-c}(D)g_d(D) \\
+ \; & g_{1-a}(D)g_{1-b}(D)g_c(D)g_d(D) + g_{1-a}(D)g_{1-b}(D)g_{1-c}(D)g_{1-d}(D). \qquad (25)
\end{aligned}
$$

For example, consider the shell count for $\mathbf{Z}^4$. From (24) we have

$$T(\mathbf{Z}^4; D) = [g_0(D)]^4 = 1 + 8\,D + 24\,D^2 + 32\,D^3 + 24\,D^4 + 48\,D^5 + \cdots,$$

which shows that the lattice $\mathbf{Z}^4$ has 1 point at the origin, 8 points on the second shell with energy 1, 24 points on the third shell with energy 2, etc. (see Table 3). In cumulative terms, there are 33 points through the first three shells. By deleting the point at the origin we have $\mathcal{M} = 32$ points whose average energy is

$$\mathcal{E} = \frac{8 \times 1. + 24 \times 2.}{32} = 1.75.$$

Since $d_{\min} = 1$, we have $\Lambda = 2.86$, while $R^* = 5/4 = 1.25$.

A design with $\mathcal{M} = 64$ is obtained by retaining the first four shells and removing the point at the origin. The average energy is now

$$\mathcal{E} = \frac{8 \times 1. + 24 \times 2. + 32 \times 3.}{64} = 2.375,$$

which gives $\Lambda = 2.53$ while $R^* = 6/4 = 1.5$.

As for the lattice D_4, with zero offset the shell populations do not readily match a power of 2. Its enumerator is

$$\begin{aligned} T(D_4; D) &= g_0^4(D) + 6\,g_0^2(D)g_1^2(D) + g_1^4(D) \\ &= 1 + 24\,D^2 + 24\,D^4 + 96\,D^6 + 24\,D^8 + 144\,D^{10} + \cdots \end{aligned} \qquad (26)$$

(see Table 4).

Shell #	Energy of shell	Signals on shell	Signals within shell
1	0.	1	1
2	2	24	25
3	4	24	49
4	6	96	145
5	8	24	169
6	10	144	313

Table 4: Shell count for D_4.

With an offset of $(0.5, 0.5, 0, 0)$ applied to D_4 we have the enumerator

$$T(D_4 + (0.5, 0.5, 0, 0); D) = 2\,g_{0.5}^2(D)[g_0(D) + g_1(D)]^2$$

(see Table 5). We find 64 points in its first five shells, a design listed in [24]. It provides

$$\mathcal{E} = \frac{2 \times 0.5 + 8 \times 1.5 + 12 \times 2.5 + 16 \times 3.5 + 26 \times 4.5}{64} = 3.375,$$

so that

$$\Lambda = \frac{2}{3.375}6 = 3.56$$

with $R^* = 1.5$. Notice that the value of Λ is very close to the asymptotic result.

These results are summarized in Fig. 5.

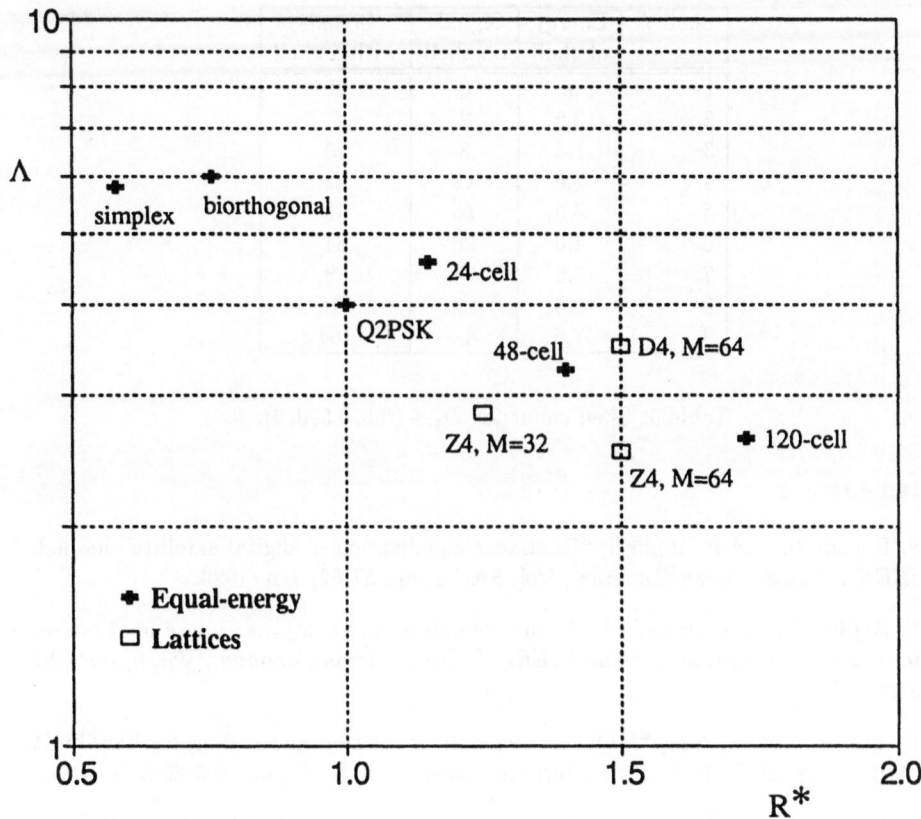

Figure 5: Power and bandwidth efficiencies of four-dimensional signal sets.

Shell #	Energy of shell	Signals on shell	Signals within shell
1	0.	0	0
2	0.5	2	2
3	1.5	8	10
4	2.5	12	22
5	3.5	16	38
6	4.5	26	64
7	5.5	24	88
8	6.5	28	116
9	7.5	48	164

Table 5: Shell count for $D_4 + (0.5, 0.5, 0, 0)$.

References

[1] S. Benedetto and E. Biglieri, "Nonlinear equalization of digital satellite channels," *IEEE J. Select. Areas Commun.*, Vol. SAC-1, pp. 57-62, Jan. 1983.

[2] E. Biglieri, S. Barberis and M. Catena, "Analysis and compensation of nonlinearities in digital transmission systems," *IEEE J. Select. Areas Commun.*, Vol. 6, pp. 42-51, Jan. 1988.

[3] E. Biglieri and M. Elia, "Multidimensional modulation and coding for bandlimited digital channels," *IEEE Trans. Inform. Theory*, vol. IT-34, pp. 803-809, July 1988.

[4] E. Biglieri and M. Elia, "Multidimensional modulation and coding for digital transmission," *Conference on Digital Processing of Signal in Communications*, Loughborough (UK), April 1985.

[5] E. Biglieri and A. Spalvieri, "Performance evaluation of coded modulation schemes based on binary lattices," *submitted for publication, 1992*.

[6] H. Brändström, *Classification of Codes for Phase and Amplitude Modulated Signals in Four Dimensional Base Space.* Telecommunication Theory, Electrical Engineering, Royal Institute of Technology, Stockholm (Sweden), Technical Report No. 105, January 1976.

[7] A. R. Calderbank and L. H. Ozarow, "Nonequiprobable signaling on the Gaussian channel," *IEEE Trans. Inform. Theory*, Vol. 36, No. 4, pp. 726-740, July 1990.

[8] P. R. Chevillat and G. Ungerböck, "Optimum FIR transmitter and receiver filters for data transmission over band-limited channels," *IEEE Trans. Commun.*, Vol. COM-30, No. 8, pp. 1909-1915, Aug. 1982

[9] A. Chouly and H. Sari, "Six-dimensional trellis-coding with QAM signal sets," *IEEE Trans. Commun.*, January 1992, to appear.

[10] J. H. Conway and N. J. A. Sloane, "Fast quantizing and decoding algorithms for lattice quantizers and codes," *IEEE Trans. Inform. Theory,* vol. IT-28, pp. 227-232, Mar. 1982

[11] J. H. Conway and N. J. A. Sloane, "A fast encoding method for lattice codes and quantizers," *IEEE Trans. Inform. Theory,* vol. IT-29, pp. 820-824, Nov. 1983.

[12] J. H. Conway and N. J. A. Sloane, *Sphere Packings, Lattices and Groups.* New York: Springer Verlag, 1988.

[13] R. deBuda, "Some optimal codes have structure," *IEEE J. Select. Areas Commun.,* Vol. SAC-7, pp. 893-899, August 1989.

[14] G. D. Forney, Jr., "Coset codes – Part I: Introduction and geometrical classification," *IEEE Trans. Inform. Theory,* Vol. 34, No. 5, pp. 1123-1151, September 1988

[15] G. D. Forney, Jr., and L.-F. Wei, "Multidimensional constellations — Part I: Introduction, figures of merit, and generalized cross constellations," *IEEE J. Select. Areas in Commun.,* Vol. 7, No. 6, pp. 877-892, Aug. 1990

[16] G. D. Forney, Jr., "Multidimensional constellations — Part II: Voronoi constellations," *IEEE J. on Select. Areas in Commun.,* Vol. 7, No. 6, pp. 941-958, Aug. 1990.

[17] A. Gersho and V. B. Lawrence, "Multidimensional signal constellations for voiceband data transmission," *IEEE J. Select. Areas Commun.,* vol. SAC-2, No.5, pp. 687-702, September 1984.

[18] G. Karam and H. Sari, "A data predistortion technique with memory for QAM radio systems," *IEEE Trans. Commun.,* Vol. 39, No. 2, pp. 336-344, February 1991.

[19] S. Pupolin and L. J. Greenstein, "Performance analysis of digital radio links with nonlinear transmit amplifiers," *IEEE J. Select. Areas Commun.,* vol. SAC-5, pp. 534-546, April 1987.

[20] D. Saha and T. G. Birdsall, "Quadrature-quadrature phase-shift keying," *IEEE Trans. Commun.,* Vol. 37, No. 5, pp. 437-448, May 1989.

[21] C. E. Shannon, "Communication in the presence of noise," *Proc. IRE,* vol.37, pp.10-21, January 1949

[22] D. Slepian, "Bounds on communications," *B.S.T.J.,* pp. 681-707, 1963

[23] M. Visintin, E. Biglieri, and V. Castellani, "Four-dimensional signaling for bandlimited channels," *IEEE Trans. Commun.,* to be published, 1992.

[24] G. R. Welti and S. L. Lee, "Digital transmission with coherent four-dimensional modulation," *IEEE Trans. Inform. Theory,* vol. IT-20, No. 4, pp. 397-402, July 1974.

[25] S. G. Wilson and H. A. Sleeper, *Four-Dimensional Modulation and Coding: An Alternate to Frequency Reuse.* Technical Report, Communications System Laboratory, University of Virginia, September 1983.

[26] S. G. Wilson, H. A. Sleeper, and N. K. Srinath, "Four-dimensional modulation and coding: An alternate to frequency reuse," *Proceedings of ICC'84*, Amsterdam, The Netherlands, May 1984, pp. 919-923

[27] L. Zetterberg and H. Brändström, "Codes for combined phase and amplitude modulated signals in four-dimensional space," *IEEE Trans. Commun.*, vol. COM-25, pp. 943-950, September 1977.

GENERALIZATIONS OF THE VITERBI ALGORITHM WITH APPLICATIONS IN RADIO SYSTEMS

Carl-Erik W. Sundberg

AT&T Bell Laboratories
Signal Processing Research Department
Murray Hill, New Jersey 07974, USA

ABSTRACT

In this paper we will give a brief overview of recent work on generalizations of the classic Viterbi Algorithm (VA) in different types of communication systems which include concatenated coding schemes. We will illustrate the usefulness of the algorithms by giving applications to speech and data transmission in radio systems. Mainly two classes of algorithms will be considered, namely list output VA or LVA and soft symbol output VA or SOVA. The LVA gives a list of the most likely output sequences while the SOVA gives the most likely sequence appended by output symbol reliability information. The additional list or soft symbol information is then processed by the next decoding stage. We will show how gains in power and/or bandwidth are obtained with list and soft output VA over schemes with a conventional hard sequence output, at the expense of increased signal processing cost. In applications where power and bandwidth are limited resources, this seems to be a reasonable path for future systems, since signal processing cost is expected to fall.

1. INTRODUCTION

The Viterbi algorithm (VA) [1-2] performs efficient maximum likelihood decoding of finite state signals observed in noise. In many situations, the state space is either too large to search or only a part of the signal is generated by a finite state machine. Five examples of such communication systems are shown in Figure 1. The first one is a concatenated coding system consisting of an outer error detecting block code (encoder 1) and an inner error correcting convolutional code (encoder 2). The second one is an outer speech coder followed by an inner error correcting convolutional code. In the former, the joint state space can be defined but it may be too huge to search exhaustively with one combined decoder, while in the latter, the combined source-channel coder state space is not even precisely defined. Conventionally, the inner code is decoded first, assuming that the input to the inner code is independent and identically distributed data, followed by outer decoding. The inner decoder is normally based on the Viterbi algorithm (VA) which searches for the best path through a trellis that is defined by the inner code. Considerable improvement in performance is obtained over this conventional decoding approach when the knowledge of $L > 1$ best paths through this trellis is utilized during subsequent processing. The algorithms we will consider in this paper belong to the class of generalized Viterbi algorithms (GVA). We use the term GVA in a broad sense to mean that the decoder produces more information than merely releasing the globally best path. For list output Viterbi algorithm we use the term LVA.

In Figure 1 we also show three cases (3-5) of concatenated coding schemes where soft symbol output Viterbi algorithms are useful. Interleaving is assumed but not shown explicitly in cases 3-5 in Figure 1. In cases 3 and 5 we have two types of classic concatenated coding schemes while in case 4, the inner "coder" is an intersymbol interference channel. We can see from Figure 1, that both LVA and SOVA are of obvious interest for noisy radio channels.

Various generalizations of the VA have appeared in the literature. Forney [3] considered a list-of-2 maximum likelihood decoder for the purpose of analyzing sequential decoding algorithms [4]. Yamamoto and Itoh (Y-I) [5] proposed an ARQ algorithm where the decoder requests for a frame repeat whenever the best path into every state at some trellis level is "too close" to the second best candidate into every state. However they do not explicitly make use of the second best candidate. More recently, Hashimoto [6] has proposed a list type reduced-constraint generalization of the Viterbi algorithm which contains the Viterbi and the M-algorithm [7] as special cases. The purpose of this algorithm is to keep the decoding complexity to be no more than that of the Viterbi algorithm and to avoid error propagation due to reduced state decoding. Again, no explicit use of the survivors other than the best is made after decoding. The Y-I algorithm has been successfully employed in a concatenated coding scheme by Deng and Costello [8] (case 5). The inner decoder is a convolutional code with the Y-I decoding algorithm and the outer decoder is errors and erasures decoder where the symbol erasure information is supplied by the Y-I algorithm. A soft output Viterbi algorithm which gives analog reliability information associated with each decoded symbol from the VA as well as a reduced complexity algorithm motivated by a maximum a-posterior (MAP) decoding rule have been proposed by Hagenauer, Hoeher and Huber [9],[10]. Such algorithms are typically used in the inner decoding stage. Combined with interleaver/de-interleaver pair, these two form powerful decoding techniques for the decoding of trellis coded data over frequency-selective multi-path fading channels.

Here, we concentrate on two list Viterbi decoding algorithms (LVA) that produce a rank ordered list of the L globally best candidates after a trellis search [11]-[14]. The two algorithms are (i) a parallel LVA that simultaneously produces the L best candidates and (ii) a serial LVA that iteratively produces the jth best candidate based on the knowledge of the previously found $j - 1$ candidates. We consider the application of this algorithm to a concatenated communication system consisting of an inner forward error correction (FEC) code and an outer (ideal) error detecting code. Later in the text, we briefly consider the application of this algorithm to other concatenated communication systems.

2. LIST VITERBI DECODING ALGORITHMS

In this section we present the parallel and serial LVA [11]. For the parallel algorithm, the task of identifying the L best candidates is achieved in one pass through the trellis while the serial algorithm achieves the same result by L successive passes through a trellis. We begin by summarizing the Viterbi algorithm which also establishes the notations to be used subsequently.

An N state fully connected trellis is shown in Figure 2. (Some of the connections are non-existent for rates less than $\log_2 N$ bits/state transition.) The total number of sequences is $N^2 M$ where M is the total number of such trellis sections. The cost (metric) associated with moving from state i to state j at time t is given by $c_m(i,j)$ (where $c_m(i,j) = \infty$ if i and j are not connected). The problem is to find the best state sequence (with minimum cost) through the trellis, starting from, for example, state 1 (at time 0) and ending at state 1 (at time M). Thus in the framework of finite state channel codes we deal mainly with terminated codes. Let the minimum cost to reach state j at time t from the known starting state be $\phi_t(j)$. Let the history of the best path be stored in the array $\xi_t(j)$. At time t, $\xi_t(j)$ is the state occupied by the best path at time $t-1$. The Viterbi algorithm can be summarized in the following steps 1. Initialization. 2. Recursion. 3. Termination. 4. Path Backtracking.

The parallel LVA finds the L best paths simultaneously. We begin by explaining the $L = 2$ best path search.

Theorem: *To release the two globally best paths, it is necessary and sufficient to find and retain the two best paths into each state at every time instant.*

Proof: The necessary condition follows from the fact that the globally second best path is distinct from the globally best path over at least a portion of the trellis. Ultimately it merges with the best path at some time. Thus, if we do not find the second best path into the state where this merging happens (and which we do not know apriori) then the globally second best path is not known at all. The sufficient condition follows from the fact that the third best candidate (or lower) cannot be a survivor for being the second best at any state since there exists a candidate, namely the second best with a better metric. This concludes the proof.

Along similar lines, it can be shown that in order to find the globally L best candidates, it is necessary and sufficient to find and retain the L best candidates into each state at every level. The recursion is illustrated in Figure 3 where at each state and at every time, the NL accumulated costs are computed and the L smallest accumulated cost paths along with their costs are stored. This algorithm requires maintaining a cost array of $\dot{N}L$ accumulated costs and a state array of $NL \times M$ which stores the path history for each time instant. It is easy to modify the parallel algorithm to operate in a continuous transmission mode rather than in a block mode by maintaining the L best paths into each state and by releasing the L best symbols (after tracking back D_p symbols back where D_p is the decoding depth) at each instant corresponding to the best among the NL survivors, the second best among the NL survivors etc.

The *serial algorithm* finds the L most likely candidates, one at a time, beginning with the most likely path. The main benefit of this algorithm is that the ℓth best candidate is computed only when the previously found $\ell - 1$ candidates are determined to be in "error". This avoids many of the unwanted computations of the parallel algorithm. Details of the serial algorithm are given in [11]. We illustrate the serial algorithm for finding the 2nd and 3rd best and then present the general algorithm. The best path is assumed to be already found by the Viterbi algorithm. It is convenient to retain all the computations performed by the Viterbi algorithm including the cost associated with each locally best (partial) state sequence.

We make use of the fact that the globally 2nd best path after leaving the best path at some instant, merges with the best path at a later instant and never diverges again. This is because any subsequent divergence will result in a higher cost path. In order to release the second best path, the time instant at which the last best merge happened should be recorded as well as the corresponding predecessor state. Supposing this time instant is τ and the predecessor state is $\ell^{(\tau-1)}$. Then, the globally second best state sequence is the best path from state 1 (at time 0) to state $\ell^{(\tau-1)}$. The remainder is the same as the globally best state sequence. We note that, unlike the parallel algorithm, it is not necessary to keep track of $2N$ accumulated costs and $2N$ path histories.

The third best state sequence is found by making use of the fact that it (and in general, the kth best path) merges with the best path at some time instant for the final time after diverging at some time in the past. We note that unlike the second best path, we do not preclude multiple divergence and merges with the best state sequence. However, the 3rd best path has the same topology as that of the 2nd best path in that once it merges with the best path, it never diverges again.

We now summarize the serial algorithm. We also assume that the globally best sequence has been found by the Viterbi algorithm and that the locally best path into each state at every time instant is known along with its associated cost. Let the best state sequence be $(1, i^{(1)}, i^{(2)}, \ldots, i^{(j)}, \ldots, i^{(M-1)}, 1)$.

Serial List Output Viterbi Algorithm:

Initialization:

    a. Initialize the number of candidates found to $\ell = 1$.

    b. Set up a path matrix of size $L \times M$, to store the state sequences, where L corresponds to the maximum number of best state sequences that are to be found, and M is the length of the state sequence.

    c. Form a "merge" count array of size M, where the jth element C_j is the number of paths, among the previously found globally best paths that finally merge with the globally best path at time j. We initialize C_j to 1 for all j.

Recursion:

    a. At time j, find the path that finally merges at time j, and one that is in contention for being the ℓth best candidate. This candidate is the $C_{j+1}{}^{\text{th}}$ best path from time instant 0 (and state 1) to time instant j (and state M). The cost of this candidate is added to the cost of the globally best path over the remainder of the trellis and is compared to the cost of the surviving candidate. The lowest of the two remains in contention.

    b. Update the $C_j{}^{\text{th}}$ entry of the merge count array, i.e., $C_j = C_j + 1$ for that time instant j when the final merge of the ℓth best path with the globally best path happens.

    c. Increment ℓ by 1.

    d. Loop back to (a) until $\ell = L$.

Note: Further computational savings can be realized by storing the currently available L best candidates in a stack. Let us assume that the ℓth best path is to be found. Then, it is clear that the ℓth globally best candidate is at least the ℓth best candidate in the stack. It can also be verified that the only other possibility is that the ℓth candidate is the 2nd best path to the $(\ell - 1)^{\text{th}}$ globally best path. This candidate is compared to the ℓth candidate in the stack and the lower of the two is the globally ℓ^{th} best path. The stack is continuously updated in the process of finding the ℓth best candidate. A tree-trellis search algorithm similar to this final version of the serial algorithm (i.e., using a stack) has been proposed by Soong and Huang [15] for speech recognition applications [16]. They use the forward trellis search using the VA for finding the locally best candidate into each state at every instant followed by a single backward tree search (backwards stack) to find the remaining $L - 1$ best paths. A fixed L has to be used for the backward tree search.

It is clear that the parallel algorithm requires L times more storage and computational costs than that of the VA. On the other hand, using the final version of the serial algorithm which stores all the intermediate computations, in order to find the ℓth best path, it is sufficient to find the 2nd best path to the $\ell - 1^{th}$ globally best path. This requires the evaluation of M path metrics. If all the intermediate computations are stored then at each instant about N metric additions and N comparisons are required [11]. Thus the total cost to find the ℓth best path is NM additions and comparisons. Some additional cost is incurred in inserting a newly found candidate into the stack if its cost is lower than that of any candidate in the stack. In any event, the computational cost is much lower than that of the parallel list Viterbi algorithm. Also, in applications where the list size varies, the *average* computational cost for the serial algorithm may be much smaller than that of the parallel algorithm where the list size is fixed at the maximum. The serial algorithm finds the ℓth best only if the $\ell - 1^{th}$ best is in "error". On the other hand, the parallel LVA has to find the L best all the time. The storage requirement to store the accumulated cost into each state at every instant is about M times higher than the VA.

3. APPLICATIONS OF LVA

Here, the outer code is an error detecting code and the inner code is a convolutional code. Conventionally, the inner decoder based on the VA releases the best decoded sequence and the outer decoder performs error detection on this decoded sequence. If an error is detected then the outer decoder, for example, may request the transmitter to retransmit. In this case the inner decoder has to perform a second Viterbi decoding etc. Here we propose that the decoder based on the LVA releases the L best candidates and the outer decoder (assumed ideal in the performance evaluation below) selects the correct one from the L best if it exists. The serial algorithm is ideally suited for this purpose. This is because the outer decoder performs error detection on the ℓth best candidate, $\ell = 1, \ldots, L$ and requests the inner decoder for the $\ell + 1^{th}$ best candidate only if the ℓth best is found to be in error. A decoding failure is declared if $\ell = L$. We note that using the serial algorithm, the task of finding the 2nd best, the third best etc., requires lower computational effort than performing a second Viterbi decoding as required in the conventional approach. Figure 4 shows the decoder operation for this concatenated coding system. We first evaluate the asymptotic error performance of this coding system in the presence of noise and fading. In particular, we are interested in the probability of the correct candidate not being among the L best. The results for the Gaussian channel are derived for any L and for any code (including the inner code being a coded modulation) while for the Rayleigh fading channel the results are derived for binary codes and $L = 2$.

The probability of incorrect decoding is evaluated for the additive white Gaussian noise channel when the LVA is used for decoding [11]. This is the probability that the correct candidate is not in the list of the L best candidates. Our analysis makes use of signal space geometry. The results show that the worst case asymptotic coding gain is independent of the code, its rate, and the modulation scheme. It provides an indication of the gain that can be achieved by practical codes or modulation schemes at high channel SNRs. We will present simulation results with specific codes to show that practical gains are close to the theoretical predictions.

When $L = 2$, we are interested in finding the probability that the correct data sequence is not among the list of the two best decoder outputs. We now have three candidates $s(\underline{a})$, $s(\underline{b})$, and $s(\underline{c})$ corresponding to data sequences $\underline{a}$, $\underline{b}$ and $\underline{c}$. These three signal points form a triangle with the edges of the triangle being at least $D_{\min}$ in length where $D_{\min}$ is the minimum Euclidean distance in signal space. The closest point in the region of error has the Euclidean distance D_{eq} from $s(\underline{a})$. This distance to $s(\underline{a})$ [11] will dominate the error probability at large channel signal-to-noise ratios. D_{eq} takes on its lowest value when all the three points $s(\underline{a})$, $s(\underline{b})$ and $s(\underline{c})$ are at $D_{\min}$ from each other. The asymptotic coding gain G of the LVA ($L = 2$) over the VA ($L = 1$) is then given by

$$10 \ \log_{10}(G) = 10 \ \log_{10} \left(\frac{D_{eq}}{D_{\min}/2} \right)^2 = 10 \ \log_{10} \left(\frac{4}{3} \right)$$

which is 1.25 dB. When $L = 3$, an error occurs whenever the correct candidate is not on the list of the three best decoded messages. The worst case signal space geometry that corresponds to the least gain is a tetrahedral packing. Here, every signal point is equidistant form the other three. The coding gain can then be evaluated as $10 \log_{10}(G) = 10 \log_{10}(3/2)$ which is 1.76 dB. For any L, we are interested in finding the tightest packing of $L + 1$ signal points in L dimensions, and this is the simplex [11] where every signal point is equidistant, and $D_{\min}$, from each of the L points. The reason for this follows from the examples above. We want to know the most likely event in which a GVA can output the globally L best candidates and not have the correct candidate on the list. In [11] we show that the worst case asymptotic gain for the GVA with L outputs over the VA as

$$10 \log_{10}(G) = 10 \log_{10} \left(\frac{2L}{L + 1} \right) . \tag{1}$$

The gains with $L = 4$, 8 and 16 are 2.04 dB, 2.50 dB and 2.75 dB respectively. The (small) loss due to the rate of the error detecting code is not taken into account in (1).

The worst case asymptotic gains presented here are somewhat optimistic for intermediate and low channel signal-to-noise ratios. The actual gain is normally somewhat smaller when the number of set of L nearest neighbors is taken into account. What seems practical to achieve with a list size of $L = 2$ and 3 are gains of about 1.0 dB and 1.5 dB respectively as shown by simulations [11].

We have performed a series of simulations for rate a $R = 1/2$, code with memory $\nu = 4$ [11]. P_{BL} is the probability that the correct alternative is not among the L best produced by the LVA. The block error probabilities P_{BL}, $L = 1, 2$ and 3 have been simulated for the Gaussian (AWGN) channel. Results in Figure 5 indicate that a gain of about 1 dB is gained for a $R = 1/2$ code with block size of 512 information bits when $L = 2$, and about 1.25 dB when $L = 3$. The channel signal-to-noise ratio used in Figure 5 is E_c/N_0 where E_c is the energy per channel symbol. The energy per information bit E_b is given by $RE_b = E_c$ where R is the code rate.

We also consider the decoding of coded data symbols subjected to independent Rayleigh fades from symbol to symbol and corrupted by additive white Gaussian noise. We assume a Rayleigh fading channel that permits coherent demodulation (ideal recovery of carrier phase) of binary PSK. Interleaving over many symbols (ideally infinite) is assumed to justify the assumption of independent fading from symbol to symbol.

On a Rayleigh fading channel, the parameter that significantly affects the asymptotic error performance of a communication system is the time diversity of the system which is defined as the minimum number of symbols in which any two transmitted signals differ. When the channel code is binary, and the modulator is BPSK, the diversity of the system is the free Hamming distance D_{free} of the code. Provided the fade is independent from symbol to symbol, the asymptotic error probability with soft demodulation and Viterbi decoding is well approximated by

$$P_e \approx \text{const.} \left(\frac{1}{\overline{\text{SNR}}}\right)^D \tag{2}$$

where $D = D_{free}$ and where $\overline{\text{SNR}}$ is the average receiver channel signal-to-noise ratio. The simulations to be presented will show that the effect of using the LVA for decoding results in an increase in diversity. For example, the largest D_{free} of memory $\nu = 4$, rate $R = 1/2$ and $R = 1/3$ convolutional codes are $D_{free} = 7$ and 10 respectively. We will show that when the $R = 1/2$ code is decoded by the LVA with $L = 2$, the error performance that is obtained behaves as if the D_{free} is about 11. Thus, use of the LVA with $L = 2$, at the expense of a small increase in receiver complexity, results in a considerable savings of redundancy. Note that the increase in receiver complexity when $L = 2$ is rather small using the serial algorithm. The computation of the effective increase in diversity for binary codes with BPSK modulation will now be carried out.

For the $L = 2$, LVA case, to find out the worst case when a transmitted codeword will not be on the list of the best two candidates from the decoder output, we will again consider three codewords $s(\underline{a})$, $s(\underline{b})$, and $s(\underline{c})$. However, instead of considering an ideal Rayleigh fading channel, we consider a simplified model where if the channel is in a fade, the received symbol is assumed to be an erasure. Otherwise, the received symbol is assumed to be demodulated perfectly. For this model, we can ask how many erasures can be made and have the correct candidate in the list of two with $L = 2$, LVA. Clearly when $L = 1$, $D_{free} - 1$ erasures can be made (leading to a diversity of $D = D_{free}$). Each of the codewords $s(\underline{a})$, $s(\underline{b})$ and $s(\underline{c})$ differ in at least D_{free} positions from the other two. First assume that D_{free} is an even number. Without loss of generality, let the transmitted codeword $s(\underline{a})$ be the all zero codeword. Let the second codeword $s(\underline{b})$ be different from the all zero codeword in exactly D_{free} positions. The third codeword $s(\underline{c})$ is also different from $s(\underline{a})$ and $s(\underline{b})$ in D_{free} positions. Let it be different from $s(\underline{a})$ and $s(\underline{b})$ in d ($< D_{free}$) positions where the first two do not differ. Then in order for $s(\underline{c})$ to differ from $s(\underline{a})$ in exactly D_{free} positions, it should have $D_{free} - d$ 1's and d 0's in those positions where $s(\underline{a})$ and $s(\underline{b})$ differ. Now, $s(\underline{c})$ and $s(\underline{b})$ differ in $2d$ positions, and since the distance between them is D_{free}, d is equal to half of D_{free}. Thus the minimum number of positions in which both $s(\underline{b})$ and $s(\underline{c})$ differ from $s(\underline{a})$ is $D_{free} + d$ which is equal to $(3/2) D_{free}$. Thus if $s(\underline{b})$ and $s(\underline{c})$ are to be selected over $s(\underline{a})$, then there should be at least $(3/2) D_{free} - 1$ erasures. The effective diversity is thus $(3/2) D_{free}$. The above example is given for an even D_{free}. When D_{free} is an odd number, the effective increase in diversity is at least $(D_{free} + 1)/2$. For example, for a code with $D_{free} = 3$, the value of D with $L = 2$ is 5. For a code with $D_{free} = 10$, the value of D with $L = 2$ is 15. Note that an increase of D quickly yields significant gains in channel signal-to-noise ratio and in error probability. The largest relative gains occur for small values of D_{free}. This is confirmed by the simulations in [11]. These results above are obtained assuming an ideal erasure channel model.

We have simulated the performance of a rate $R = 1/2$ convolutional code with memory $\nu = 4$ on a Rayleigh fading channel with BPSK modulation. The demodulation is assumed to be ideally coherent, and the fading is assumed to be independent from symbol to symbol. We have not used any channel state information (CSI) (the fade values) at the decoder [11]. The metric used for decoding is the correlation metric (soft decision), the same as for the AWGN channel. The simulation results in Figure 6 for the $R = 1/2$ code show a linear relationship between P_{BL} and the $\overline{SNR}$. By calculating the slope, one can evaluate the diversity. It can be seen in Figure 6 that the diversity is about 5 when $L = 1$, and is about 9 when $L = 2$. Note that the diversity at large SNRs and with perfect CSI is 7 (value of D_{free}) when $L = 1$ and about 11 when $L = 2$ (based on the analysis above). These values are naturally lower for small SNRs and without CSI as the graphs indicate. Note that the gain with the LVA ($L = 3$) is about 3 dB in $\bar{E}_c/N_0$ over the VA ($L = 1$) case at $P_{BL} = 10^{-4}$.

Combinations of forward error correction (FEC) and automatic repeat request (ARQ) systems are often referred to as hybrid ARQ schemes [11]. On very noisy channels like fading mobile radio channels, powerful FEC is needed. The most flexible and robust hybrid ARQ schemes uses inner rate compatible punctured convolutional (RCPC) codes. These codes can be decoded using soft Viterbi decoding. The main advantage of the RCPC codes over other

FECs is their incremental redundancy transmission feature using the concept of rate compatible puncturing. The hybrid FEC/ARQ scheme with RCPC codes and Viterbi decoding works as follows. A block of N_i data bits are augmented with N_c cyclic redundancy check (CRC) bits using an outer block code. For the purpose of analysis and simulations, we will assume that the probability of undetected error for this code is zero. ν known information bits are added to the $N_i + N_c$ bits before being encoded by the inner RCPC code. These ν bits are used to terminate the trellis in a known state. This block of $N_T = N_i + N_c + \nu$ bits are encoded by a family of RCPC codes with memory ν, and rates from 1 to $1/n$. In our examples, we will use a value of $n = 3$, and puncturing period $p = 8$. The possible rate are $p/(p + \lambda)$, $\lambda = 0, 1, 2, \ldots, (n-1)p$. The parameter λ is called the level of puncturing. The information bits are first encoded by the rate $1/n$ convolutional code. The output of the convolutional code is then punctured according to a rate compatible puncturing rule which for a given value of λ is in the form of a puncturing table $\mathbf{a}(\lambda)$ [11].

In a typical ARQ scenario, a subset of all possible λ is used, e.g., $\lambda = 1, 2, 4, 8$ and 16. The transmitter starts with the highest code rate possible, and will continue to transmit additional punctured bits corresponding to successively lower rate codes until it receives positive acknowledgement from the receiver. We assume an error free feedback channel. Two parameters are used to characterize the performance of the hybrid FEC/ARQ system are the throughput S and the probability of a decoding failure P_F. The throughput S is the average number of received accepted bits over the average number of transmitted bits. There is a nonzero probability that correct decoding cannot be achieved at the lowest code rate. This is the probability of failure to decode, P_F. This quantity can be reduced by decreasing the code rate $1/n$ or through the process of code combining [11]. Both these methods will reduce the system throughput. We intend to use the LVA instead of the VA to reduce P_F and *simultaneously increase* the throughput (and implicitly decrease the overall delay) [11].

We have simulated the complete hybrid FEC/ARQ system for the $\nu = 4$ RCPC codes with the LVA ($L = 3$). The puncturing period is $p = 8$. For reference, we have also simulated the same system using a conventional Viterbi decoder (VA). The results are shown in Figure 7 for the ideally interleaved Rayleigh fading channel with BPSK modulation and soft decision decoding. Ideal channel state information is also used. The RCPC codes chosen have $\lambda = 1, 2, 4, 8$, and 16 giving code rates of $8/9, 8/10, 8/12, 8/16$, and $8/24$ respectively. As an outer code, we chose an hypothetical block code with N_c of 34. The information block size N_i was chosen to be 382. The relative improvement of the throughput is about 10% over the signal-to-noise ratio range. This corresponds to about 1 dB in $\bar{E}_c/N_0$. For further details, see [11].

Combined with the improvement in throughput comes the significant improvement in P_F. We did not explicitly simulate P_F for the $\nu = 4$, $R = 1/3$ code which is used as the final low rate code in the system in Figure 7. However this can be approximately estimated by comparing P_{B1} (VA) and P_{B3} (LVA, $L = 3$) in Figure 6 for the $\nu = 4$, $R = 1/2$ code. A gain in $\bar{E}_c/N_0$ of about 3 dB is achieved at $P_{BL} = 10^{-4}$.

Another natural application area for the LVA is the speech and channel coding scenario in Figure 1. Here the $L = 2$, LVA can be used together with a metric difference threshold for block error detection. The redundancy in the speech parameters is used to replace the detected block error. In a more advanced application, the LVA outputs the L best candidates and the speech redundancy is used for selection, see [11]–[14]. In the first generation digital mobile radio schemes in Europe and North America, a block code is used for error detection. An LVA (used in the data communication scenario above with fixed codes) can directly be used for further speech enhancement [11], without changing the standards.

4. SOVA AND MAP WITH APPLICATIONS

In 1971 Clark and Davis described an extended Viterbi Algorithm [17] which outputs a two level symbol reliability indicator. In recent years, several algorithms have been proposed which put out soft reliability information individually for each bit. Such algorithms are the *weighted output Viterbi algorithm* as proposed by Battail [18,19], the *soft-output Viterbi algorithm* (SOVA) as proposed by Hagenauer and Hoeher [9], and the *reliability estimation algorithm* as proposed by Huber [20]. The major idea of these algorithms is to transform path metric differences, which indicate the reliability of competing path *sequences*, into reliability values for individual *symbols*. Stored reliability values are recursively updated with each survivor extension. The weighted output Viterbi algorithm requires more updates than the SOVA. Suboptimum implementations of the latter are reported in [10].

In parallel, several promising suboptimum implementations of the symbol-by-symbol MAP algorithm have been proposed in particular for Viterbi equalization [22][23]. More details on the implementations are given in [10]. However it should be indicated here that suboptimum implementations of the symbol-by-symbol MAP algorithms are related to the Viterbi algorithm [10]. The 'classical' application for soft-output algorithms is concatenated coding. The motivation for concatenation is to break-up long block or convolutional codes into short codes, which are easily decodable.

For cases 3-5 in Figure 1, a soft symbol output decoder is required for the inner code. For case 5 a special case of erasure output decoder is used [10], while for cases 3 and 4, a soft symbol output decoder 1 is wanted for feeding soft input symbols to the next stage. One possibility is a MAP decoder, which is dealt with in [10]. Another possibility is a modified Viterbi algorithm, a soft output VA, a SOVA [10]. The key ideas here is that the decoder (1) calculates the path metric difference between paths merging at each state, (2) calculates the information symbol difference sequence for those paths and (3) updates the reliability of the symbols where those paths are different [10]. Only a certain window is observed, the so called update window.

It has been found in e.g. [10], that the coding gain for cases 3 and 4 can be improved by several dB, by using a SOVA in first stage. Both Gaussian and Rayleigh fading channels have been considered. Particularly interesting applications are classic concatenated coding for error correction (case 3) and equalization followed by error correction decoding. We note again that *interleaving* is required between the codes in cases 3–5. This causes increased delay. The LVA applications (cases 1 and 2) are sequence oriented and require no extra interleaving between the two decoding stages.

5. DISCUSSION AND CONCLUSIONS

We have given a brief overview of recent developments in generalizations of the Viterbi Algorithm. Two classes of algorithms have been discussed. One class provides the list of the L most likely output *sequences*. The other class provides soft output *symbol* information. Applications for both classes of algorithms have been given.

We have presented parallel and serial list Viterbi decoding algorithms [LVA] that produce an ordered list of $L > 1$ globally best candidates after a trellis search. Novel methods for utilizing the LVA for concatenated communication systems are described. The algorithm lends itself conveniently for use with rate compatible punctured convolutional (RCPC) codes for hybrid FEC/ARQ data transmission.

The LVA is directly applicable to combined speech and channel coding [11]. Lee and Rabiner [16] and Soong and Huang [15] have also considered the application of LVA to speech recognition.

We have demonstrated that the LVA improves *both* the throughput and the probability of failure to decode for hybrid FEC/ARQ schemes. The price paid is increased signal processing at the receiver. It is interesting to note that the use of the LVA is optional. One receiver may operate with VA while another may use $L = 2$, and yet another with $L = 8$ etc. It is also not necessary to use the LVA in every stage of decoding in the hybrid FEC/ARQ scheme that uses RCPC codes. If one were interested in solely reducing the probability of failure to decode, then the LVA needs to be used only in the last stage of decoding.

Much remains to be done in terms of theory, both for SOVAs and LVAs. When a conventional Viterbi Algorithm is used for decoding, the error performance can be bounded by use of union bound, transfer function bound technique etc. [1],[2]. For the LVA with L outputs this seems much more difficult. For example, when $L = 2$, triplets of codewords have to be considered instead of the pairwise error probabilities for the VA. Due to such mounting difficulties we settled for asymptotic error event behaviour and computer simulations in our evaluation. Some theoretical results have been obtained when the LVA ($L = 2$) is used for combined error correction and detection [14].

In the analysis and simulations above we assumed binary convolutional codes (terminated into block codes) and binary modulation. The results are more general, however. As we pointed out above, the LVA can be applied wherever the VA can be applied. For the Gaussian channel, the results above are directly applicable to any code and signal structure with arbitrary alphabets. For example, the results apply to combined channel coding and modulation, in particular to trellis-coded modulation [11] and coded and uncoded continuous phase modulations [11]. For the Rayleigh fading channel with ideal interleaving, there is a one-to-one correspondence between the Hamming distance of the binary code and the number of branches of time diversity. Similar results hold for multi-level codes that utilize component binary codes [11] to address the indices of the multilevel modulation. An interesting emerging application of the list decoding algorithm is to the problem of joint data and channel estimation [23] without using a training sequence for channel estimation (so called blind equalization).

Finally, we would like to mention a third class of generalizations of the classic Viterbi algorithm, namely efficient suboptimum decoding of tailbiting convolutional codes, see [24] and [25] for examples. In these algorithms, the convolutionally coded data is transmitted in a (short) block without the use of a known tail to end in a known state. The data is normally transmitted with the same starting and ending state which is unknown to the receiver. The optimum decoder performs S Viterbi decoding trials where S is the number of states. Only one trial is required when the starting/ending state is known. One efficient iterative algorithm which uses much fewer trails for high channel SNR is given in [25].

References

[1] A. J. Viterbi, "Error Bounds for Convolutional Codes and an Asymptotically Optimum Decoding Algorithm," IEEE Trans. on Inf. Theory, Vol. IT-13, pp. 260-269, 1967.

[2] A. J. Viterbi and J. K. Omura, "Principles of Digital Communication and Coding," McGraw-Hill, NY, 1979.

[3] G. D. Forney, Jr., "Convolutional Codes II: Maximum Likelihood Decoding," Inf. Control, 25, pp. 222-266, July 1974.

[4] G. D. Forney, Jr., "Convolutional Codes III: Sequential Decoding," Inf. Control, 25, pp. 267-297, July 1974.

[5] H. Yamamoto and K. Itoh, "Viterbi Decoding Algorithm for Convolutional Codes with Repeat Request," IEEE Trans. Inf. Theory, IT-26, pp. 540-547, September 1980.

[6] T. Hashimoto, "A List-Type Reduced-Constraint Generalization of the Viterbi Algorithm," IEEE Trans. Inf. Theory, IT-33, pp. 866-876, November 1987.

[7] J. B. Anderson and S. Mohan, "Sequential Coding Algorithms: A Survey and Cost Analysis," IEEE Trans. Commun., Vol. COM-32, pp. 169-176, Feb. 1984.

[8] R. H. Deng and D. J. Costello, Jr., "High Rate Concatenated Coding Systems Using Bandwidth Efficient Trellis Inner Codes," Dept. of Electr. and Computer Eng. Univ. of Notre Dame, Notre Dame, IN 46556, May 1987. Private Communication.

[9] J. Hagenauer and P. Hoeher, "A Viterbi Algorithm with Soft-Decision Outputs and its Applications," GLOBECOM '89, Dallas, Texas, Nov. 1989, Conf. Rec. pp. 1680-1686.

[10] J. Hagenauer, P. Hoeher and J. Huber, "Soft-Output Viterbi and Symbol-by-Symbol MAP Decoding: Algorithms and Applications," In submission to IEEE Trans. on Com.

[11] N. Seshadri and C-E. W. Sundberg, "List Viterbi Decoding Algorithms with Applications," In submission to IEEE Trans. on Com.

[12] N. Seshadri and C-E. W. Sundberg, "Performance of the Generalized Viterbi Algorithm for Hybrid FEC/ARQ Data Transmission," Conf. Rec. 24th Conf. on Information Sciences and Systems, pp. 471-476, Princeton, NJ, March 1990.

[13] W. C. Wong, N. Seshadri and C-E. W. Sundberg, "Estimation of Unreliable Packets in Sub-band Coding of Speech," Proceedings of the IEE, Part I, January 1991.

[14] N. Seshadri and C-E. W. Sundberg, "Generalized Viterbi Algorithms for Error Detection with Convolutional Codes," Conf. Rec. Globecom '89, pp. 1534-1538, Dallas, Texas, Nov. 1989. Also in submission to IEEE Trans. on Com.

[15] F. K. Soong and E. F. Huang, "A Tree-Trellis Based Fast Search Algorithm for Finding the N Best Sentence Hypotheses in Continuous Speech Recognition," Proc. IEEE Int. Conf. Acoustics, Speech and Signal Processing, 1991, pp. 705-708.

[16] C. H. Lee and L. R. Rabiner, "A Network-Based Frame Synchronous Level Building Algorithm for Connected Word Recognition," Proc. IEEE Int. Conf. Acoustics, Speech and Signal Processing, 1988, pp. 410-413.

[17] G. Clark and R. Davis, "Two Recent Applications of Error-Correcting Coding to Communications System Design," IEEE Trans. on Com., pp. 856–863, Oct. 1971.

[18] G. Battail, "Weighting the Symbols Decoded by the Viterbi Algorithm," IEEE Int. Symp. Inform. Theory., Ann Arbor, Calif., p. 141 (abstract), Oct. 1986.

[19] G. Battail, "Building Long Codes by Combination of Simple Ones, Thanks to Weighted-Output Decoding," in Proc. URSI ISSSE, Erlangen, pp. 634–637, Sept. 1989.

[20] J. Huber and A. Rueppel, "Reliability Estimation for Symbols Detected by Trellis Decoders," (in German), AEU, Vol. 44, pp. 8–21, Jan./Feb. 1990.

[21] W. Koch and A. Baier, "Optimum and Sub-Optimum Detection of Coded Data Disturbed by Time-Varying Intersymbol Interference," in Proc. GLOBECOM'90, San Diego, Calif., pp. 807.5.1–807.5.6, Dec. 1990.

[22] P. Hoeher, "TCM on Frequency-Selective Fading Channels: A Comparison of Soft-Output Probabilistic Equalizers," in Proc. GLOBECOM'90, San Diego, Calif., pp. 401.4.1–401.4.6., Dec. 1990.

[23] N. Seshadri, "Joint Data and Channel Estimation Using Fast Blind Trellis Search Techniques," Globecom 90, December 3-5, San Diego. Conference Record, pp. 1659-1663, also in submission to IEEE Trans. on Com.

[24] C-E. W. Sundberg and N. Seshadri, "Digital Cellular Systems for North America," GLOBECOM'90, San Diego, Dec. 1990, Conf. Rec., pp. 533–537.

[25] Q. Wang and V. K. Bhargava, "An Efficient Maximum Likelihood Decoding Algorithm for Generalized Tailbiting Convolutional Codes," IEEE Trans. on Com., pp. 875-879, August 1989.

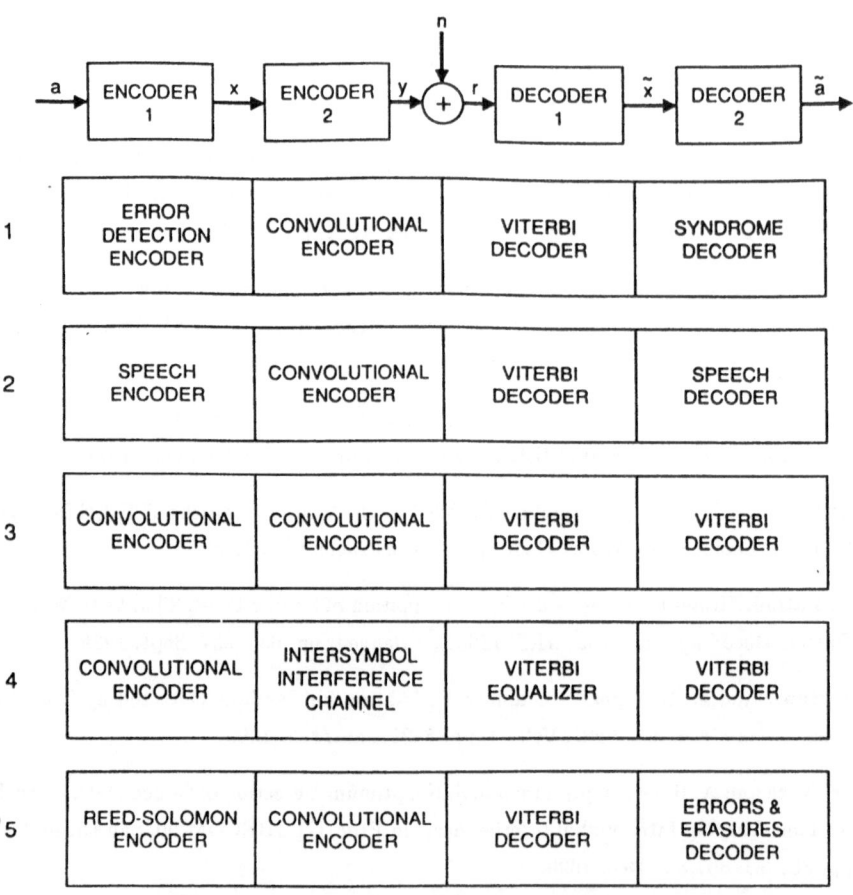

Figure 1: Examples of communication systems with concatenated decoders.

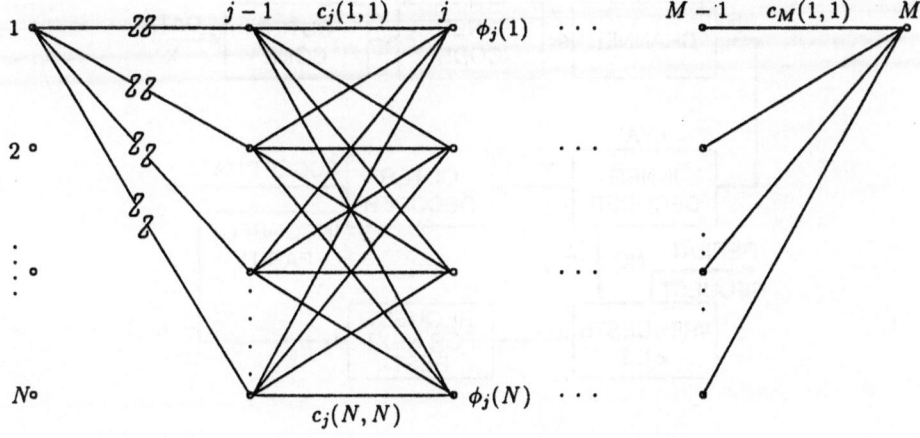

Figure 2: Fully connected trellis with N states.

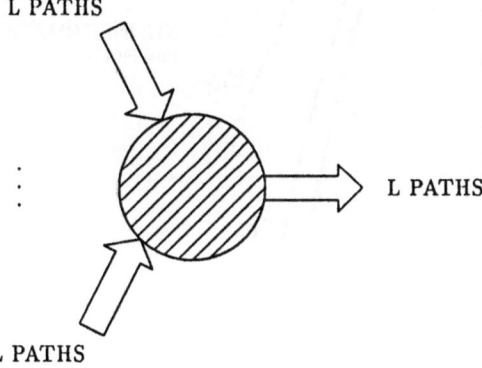

Figure 3: Dynamic programming for finding the L-best paths implemented in a parallel manner.

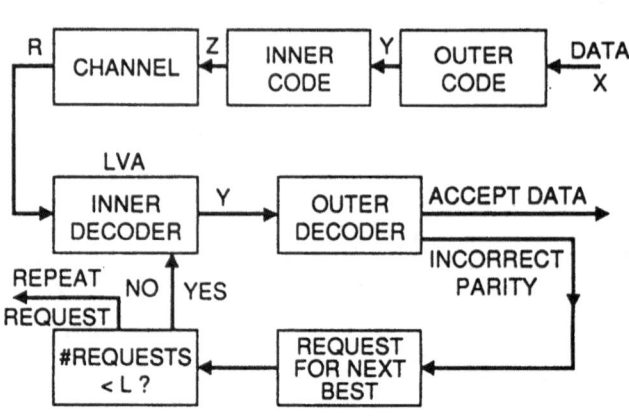

Figure 4: Block diagram of data transmission system with LVA decoder.

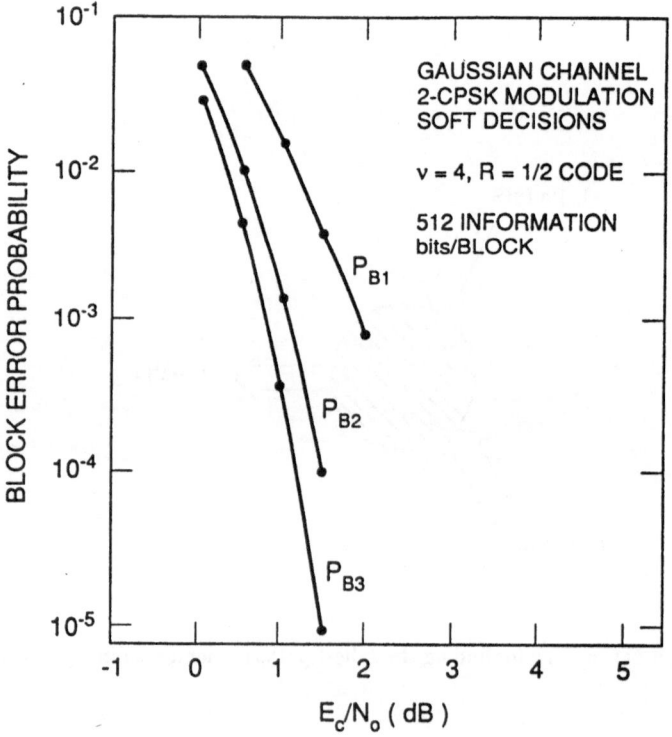

Figure 5: Block error probability up to 3-best path LVA for the Gaussian channel. An inner $R = 1/2$, $\nu = 4$ code is used. The number of information bits is 512.

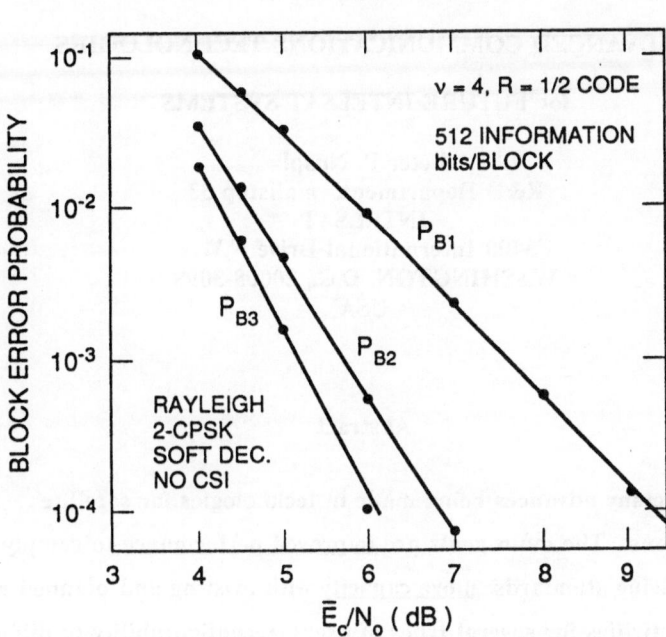

Figure 6: Block error probability for up to 3-best path LVA on the fully interleaved Rayleigh fading channel. An inner $R = 1/2$, $\nu = 4$ code is used. The number of information bits per block is 512.

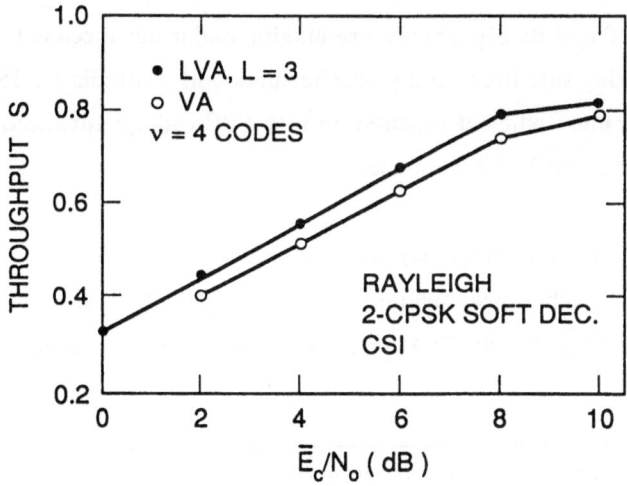

Figure 7: Throughput versus average channel $\overline{\text{SNR}}$ $\bar{E}_c/N_0$ for a hybrid FEC/ARQ scheme with LVA and VA respectively.

ADVANCED COMMUNICATIONS TECHNOLOGIES

for FUTURE INTELSAT SYSTEMS

Peter P. Nuspl
R&D Department, mailstop 33
INTELSAT
3400 International Drive N.W.
WASHINGTON, D.C., 20008-3098
USA

Abstract

There are many advances being made in technologies for satellite telecommunications. The main goals are improved performance to comply with existing and evolving standards; more capacity with existing and planned resources; increased connectivities for several types of users; reconfigurability to different and changing traffic scenarios; and better availability of the services to meet user expectations. For bus technologies, significant goals are lower mass of non-payload systems and longer life in useful orbits.

INTELSAT and its Signatories are making major advancements in the ground segment: preparing satellite-friendly architectures and protocols for ISDN, B-ISDN, SDH and ATM; more efficient modulation with FEC coding; advanced TDMA systems and enhanced VSAT networks.

In preparation for future payloads, INTELSAT has developed several on-board processing (OBP) techniques, including Satellite-Switched TDMA (SS-TDMA), Satellite-Switched FDMA (SS-FDMA), several Modulators / Demodulators (Modem),

---

The work reported here was conducted as part of INTELSAT's Research and Development Program. Views do not necessarily represent the views of INTELSAT.

a Multi-Carrier Demultiplexer (MCD) and a Demodulator (MCDD), a BaseBand
Processor (BBP), and an optical transceiver. Some proof-of-concept hardware and
software were tested in 1990 and 1991 in the INTELSAT Technical Laboratories; the
MCDD and Time-Division Multiplexing (TDM) developments for engineering models
are in progress.

The most noteworthy results are 6 x 6 CONNECTIVITY (and up to 16 x 16)
using switch matrices, reductions of 10 dB in EIRP in uplinks and more CAPACITY
(about double) or improved PERFORMANCE from the use of regeneration in the
satellite, and much higher CAPACITY when TDM is used in the downlinks. The
highest impact is likely to be on VSAT networks which will be MESH CONNECTED,
since regeneration, decoding, switching and multiplexing in the satellite perform most
essential central functions.

1. INTRODUCTION

This paper begins with discussions of the reasons for advancements
in future satellite systems and closely associates these with service requirements.
Next, some of the technologies under development for the ground and space segments
are identified and described. Some trends in satellite communications are discussed.

2. WHY ADVANCED TECHNOLOGIES ?

There are several major reasons for seeking advancements in
technologies for communications satellites and earth stations: improved performance
to comply with existing and evolving standards; more capacity using existing and
planned resources; enhanced connectivity for several types of users; reconfigurability
to be responsive to different and changing traffic scenarios; and better availability of
the services to meet user expectations. For satellite bus technologies, significant
goals are lower mass of these subsystems and longer life in useful orbits.

Clearly a significant driving force for such advances is that communications satellites are facing dramatic competition with alternative (terrestrial, undersea) transmission means [1]. More sophisticated and flexible user-oriented satellite system architectures are being studied and developed to increase the efficiency, to lower the overall system cost (space and ground segments), and to meet the requirements for low-cost, smaller earth stations to directly access satellites at low-to-moderate data rates. Delay, which must not be unduly increased during signal processing, and information throughput are very significant parameters for some services. Such requirements could result in having many advanced features in satellite payloads, and implementing some suitable combinations of network protocols, access, modulation, and FEC coding schemes.

Several advanced technologies are briefly described, and two of them - trellis-coded modulation and multicarrier demodulation - are presented in some detail, with emphasis on operational constraints. Of course, INTELSAT Engineering & Research is also following many other developments by others around the world.

3.   SERVICE REQUIREMENTS

INTELSAT is moving rapidly to provide most of its services using digital techniques: of the several reasons for this, improved performance is a major requirement from users. Bit error ratio (BER) is the usual measure, and this is closely coupled to availability, which is expressed as a percentage of the year that the service quality is provided. Table 1 illustrates the PERFORMANCE requirements for the main existing digital services: a specified BER of 1.0E-3 or lower, for 99.98% of the time. To meet CCITT Recommendation G.82x, a goal of 1.0E-6 is our objective, for the same availability. INTELSAT IBS / IDR[1] services already use

---

[1]   IBS is the International Business Service and IDR stands for Intermediate Data Rate, INTELSAT's range of public switched telephony services.

Table 1: SERVICE REQUIREMENTS - SYSTEM PERSPECTIVE
for <u>EXISTING INTELSAT SERVICES</u>

SERVICE	PERFORMANCE	CAPACITY	COVERAGE & CONNECTIVITY

Telephony
 	
(TDMA
and
SS/TDMA)
	<u>Specifications:</u>
	  BER better than
	1.0E-03 for > 99.98%
	1.0E-06 for > 99.96%
	1.0E-10 for > 96%
		<u>Present:</u> rate-7/8 FEC
		
		<u>Possibilities:</u>
		  remove FEC
		  adaptive FEC
		  higher speed
			<u>Present:</u>
			  6-fold re-use in C bands
			
			<u>Possibilities:</u> C-to Ku bands
			          Ku-to-C bands
			  8-fold re-use in C bands
			  4-fold re-use in Ku bands

IBS
and
IDR
	<u>Goals:</u>
	  BER better than
	1.0E-06 for > 99.98%
	1.0E-08 for > 99.96%
	1.0E-10 for > 96%
	
	under consideration:
	  R-S outer codec
		<u>Present:</u> rate-1/2 FEC
		         rate-3/4 FEC
		
		under consideration:
		  rate-7/8 FEC
		  Trellis-Coded 8PSK
		
		<u>Goal:</u> better
		     spectral efficiency
			<u>Present:</u> no hemi-beams
			        in Ku bands
			
			<u>Possibilities:</u>
			  hemi-like beams
			     in Ku bands
			  hopping beams

Table 2: SERVICE REQUIREMENTS - SYSTEM PERSPECTIVE
for PLANNED INTELSAT SERVICES

SERVICE	PERFORMANCE	CAPACITY	COVERAGE & CONNECTIVITY
VSAT 64 kbit/s to 2.048 Mbit/s	Specifications: BER better than 1.0E-03 for > 99.96% 1.0E-06 for > 99.36% 1.0E-07 for > 96 %	Future: HIGHER CAPACITY for HUB to VSAT HUBLESS SYSTEM to reduce delay	WIDE AREA C and Ku bands Possibilities: C-to-Ku bands Ku-to-C bands
Digital TV	Expected: BROADCAST QUALITY BER about 1.0E-05 under degraded sky (before INTERNAL FEC)	Future: up to 5 carriers / 72 MHz	WIDE AREA and MULTI BEAMS
Digital Satellite News Gathering	Expected: selected compression; transportability	Future: several carriers in 36 MHz	WIDE AREA for DOWNLINK Ku or Ka bands for ROVING UPLINKS

convolutional coding (rates 1/2 and 3/4) with soft-decision Viterbi decoding, and Reed-Solomon (R-S) outer codes are being considered for practical additional coding gain of about 2 dB.

Table 2 indicates less stringent performance requirements for some planned services. In all Ku-band services, the required high availability (all but 2 hours of a year) controls the design of digital links. Note that new digital TV services are less demanding of the transmission links, since the TV codecs (are expected to) have additional internal FEC; compatibility of these two forms of FEC codes is a present concern.

Given the existing and planned satellites and earth stations, there is also a need for more capacity, as measured in terms of derived 64-kbit/s channels for example. The most significant recent advancement is Digital Circuit Multiplication Equipment (DCME) which uses 32-kbit/s ADPCM for voice services in "open networks"[2] with multi-destinational features [2]; such equipment is being installed and will provide five times as many voice circuits, compared to 64-kbit/s PCM. Facsimile compression techniques are also being standardized; the solution is to demodulate the voice-band signal at the earth station, send the digital stream according to the DCME protocol, and reconstruct the voice-band signal [3].

Also, for mobile and thin-route voice services, LD-CELP (low delay, code excited linear processor) at 16-kbit/s has been selected [4], and (vocoder) rates as low as 4.8 kbit/s will be used for specialized applications. For larger earth stations using TDMA, which are now bandwidth limited, removal of rate-7/8 FEC is being considered, and for IDR services which now use rate-3/4 codes, less coding could be applied in satellite links with higher power. Of course, the performance / capacity trade-offs must be examined carefully for each service. Tables 1 and 2 summarize some considerations under CAPACITY.

Communications satellites have the unique capability to provide WIDE AREA

---

[2]   In this context, an "open network" could have equipment and software from several vendors who use the specifications prepared by INTELSAT.

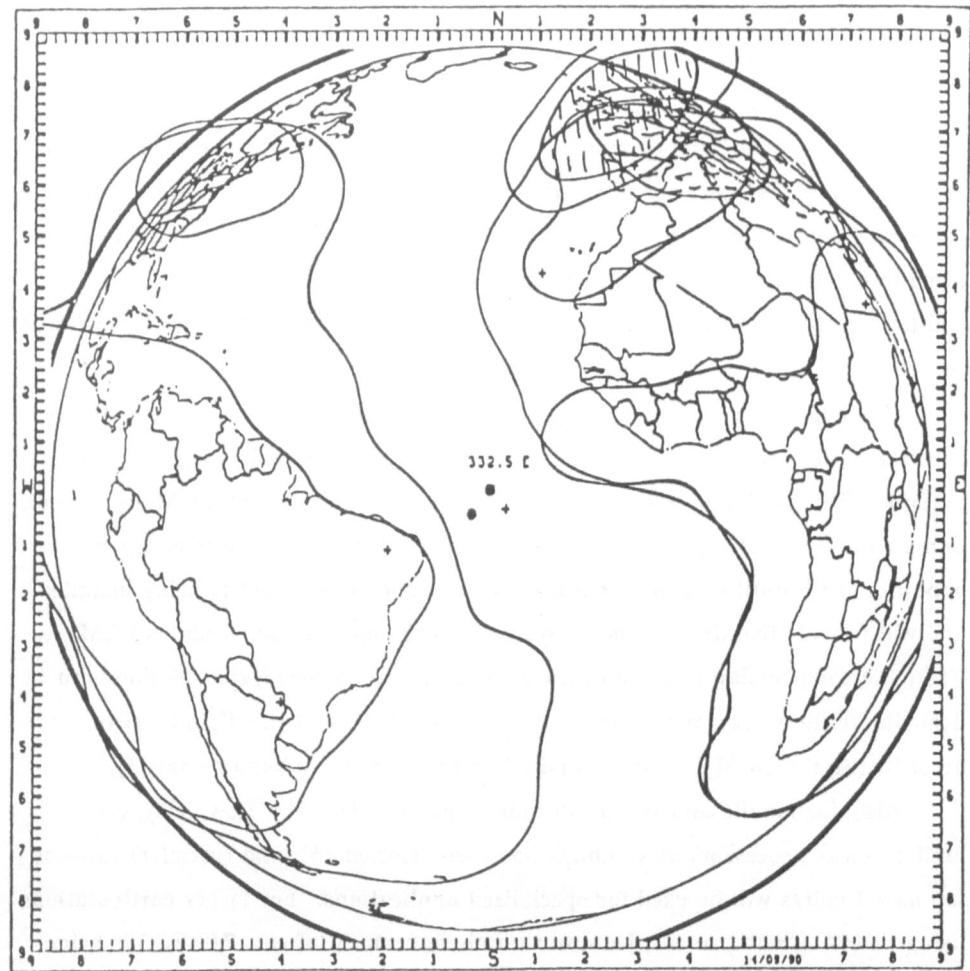

Figure 1: Transmit Beam Coverages
for INTELSAT VI at 332.5°E

COVERAGES; for illustration, Figure 1 shows the types and extent of coverages with the INTELSAT VI satellite at 332.5° East. Note the global beam, the two hemi beams and the four zone beams in C-band. The spot beams in Ku band (cross-hatched) are used to provide higher EIRP to regions with heavy demands. The re-use of frequency bands permits more capacity, but connectivity must then be provided by other means, preferably through on-board routing and switching. Tables 1 and 2 list the issues in considering COVERAGE and CONNECTIVITY along with performance and capacity.

4.   ADVANCED EARTH TERMINALS AND NETWORKS

The communications satellite industry - suppliers, operators, and users - is very adept at making major advances in the ground segment. Perhaps in the longer term, networking issues will be the most significant, and specific system improvements are also very important in the light of the above objectives. Video compression techniques have had significant improvements in the past two years, so that conventional TV would require from 6 to 10 Mbit/s and advanced TV (ATV) could use 20 to 45 Mbit/s; ATV systems with high definition are undergoing extensive tests and can also be applied in high resolution services, such as interactive graphics. Other technologies being developed for use in earth stations include advanced TDMA terminals for telephony, a new TDMA system for digital video services, protocols and transmission systems for ISDN, B-ISDN and ATM, and "satellite-efficient protocols".

n-ISDN, B-ISDN and ATM

Protocols   The satellite industry has recognized the need for satellite-friendly protocols in emerging standards and transport systems, and is actively pursuing appropriate goals [5]. As examples, for better throughput we have increased the window size in the X.25 protocols, and there is development in progress for a satellite-efficient protocol for X.75 transmissions via satellite [6].

Coded 8PSK Transmission Along with coded QPSK in existing services, the importance of combined use of FEC coding and octal PSK (8PSK) modulation has long been recognized; developments and field tests were completed under real operational conditions. Recently, some of our users have acquired coded 8PSK modems which operate at 140 Mbit/s through existing 72-MHz transponders and are intended for optical fiber restoration services. This equipment uses rate-5/6 convolutional encoding, soft-decision demodulation and Ungerboeck decoding [7].

TCM for B-ISDN Broadband ISDN (B-ISDN) has had much attention in the CCITT Study Group XVIII, other international organizations and numerous national groups. Significantly, a bit rate has been selected: 155.52 Mbit/s information rate. Other performance requirements are being defined.

Here too, satellite operators need to provide such services through existing and planned 72-MHz transponders. In an on-going project there are developments of trellis-coded modulation (TCM) with multi-dimensional signals and a concatenation of inner and outer codes for most effective corrections. The modem is 8PSK and will have a symbol rate of 62 or 66 Msymbol/s, where filters are installed for the selected rate; modem implementation margins are as high as 1.5 dB. The inner codec is to be evaluated and selected as either a rate 5/6, dimension D = 4; or rate 8/9, D = 6 code from a family of multi-D convolutional codes. It has the function of bringing the raw BER from about 2.0E-2 (worst case) to about 6.0E-4 or better, at E_b/N_o = 6.8 dB or 7.8 dB, respectively. The outer codec will use a R-S(255,239) code, which corrects up to eight symbol errors, so as to produce the objective BER of 1.0E-10 or better, at E_b/N_o = 7.1 dB or 8.1 dB, respectively. A small implementation margin is allowed for the codecs.

TCM for IDR With the objectives of more EFFICIENCY (2 bits/Hz) and more CAPACITY with existing Standard A earth stations using multi-carrier FDMA, we have another development which will apply trellis-coded modulation in IDR services at information rates up to 8 Mbit/s. With 8PSK modulation, the inner code at rate 2/3 is to be selected based on performance and implementation factors; for example, compatibility with existing coded QPSK would be an advantage. Significantly, the outer code will also be R-S(255,239).

Advanced VSAT Networks

Very Small Aperture Terminals (VSAT), ultra small terminals (USAT) and so-called microterminals are used to establish specialized networks in many parts of the world, in C and Ku bands as available. For data gathering as an example, Intelnet I and II use small terminals which work with larger hubs; there are 19 hubs and more than 1000 terminals. For many widespread organizations, point-of-sale transactions are processed at network hubs; the present network topology is a STAR. These ground-based hubs are used for link advantages, centralized control, etc. Advanced networks of VSATs should be MESH CONNECTED so that VSATs can communicate directly via satellite; there is only a single-hop delay (250 ms), the data passes through the satellite once, and essential network control functions would remain centralized at a few locations using VSATs (not large antennas). Multi-carrier regeneration and single-carrier downlinks, discussed below, will immensely assist such MESH VSAT networks.

5.   ADVANCED PAYLOADS

There are of course several types of new payloads, for example in the INTELSAT VII, VII-A and INTELSAT K satellites already contracted, the "Follow-On Satellites" planned by INTELSAT, the experimental payloads sponsored by ESA, NASDA and NASA, and conceptual payloads for beyond the year 2000. A few special advanced payload items are described below.

On-Board Processing

OBP subsystems, which can be broadly characterized as providing significant conditioning of traffic signals, are appropriate for meeting many future requirements.

Satellite switching increases the FLEXIBILITY of resource utilization. Regeneration improves link PERFORMANCE and offers alternatives to the approach of merely increasing the transmitted power from the satellite and G/T of earth stations. Under its R&D Programs, INTELSAT has developed several OBP architectures and techniques [8], and they include a microwave switch matrix for SS-TDMA (operational, and an advanced MSM), SS-FDMA channelizing filters, burst modems, MCDD, IBS / IDR, BBP and other items [9, 10, 11]. Some of these advances are good for both PERFORMANCE and CONNECTIVITY improvements. There are some which are simpler and low-risk technologies, and can be specified in the near future. For others, there are some technical problems and system issues which should be resolved before the technologies can be used on-board [12].

Microwave Switch Matrix   INTELSAT currently has two networks using SS/TDMA on INTELSAT VI satellites which each have a (dynamic) microwave switch matrix (MSM); the main benefit is CONNECTIVITY among 6 beams in C bands. The best application is in the Indian Ocean Region, where 25 uplinks of small to medium sizes are connected to 25 downlinks. The MSM consists of P.I.N. diodes as the active switching elements which work with directional couplers. Circuit paths through this MSM are somewhat variable and heavily attenuated, so that amplifiers are needed in each output. There are minimal on-board diagnostics.

Improved switching elements use graduated couplers with dual-gate FETs which have a small, steady gain and are consistent for all circuit paths. A tested engineering model has built-in redundancies and diagnostics. All technologies are space qualified or could be qualified. For dimensions up to 10 x 10 (functional) this FET-MSM is likely the preferred solution.

In comparison to similar connectivity in FDMA systems like those used in IDR, SS/TDMA systems have IMPROVED CAPACITY because the transponders can be operated at or near saturation, and exhibit more FLEXIBLE CONNECTIVITY which is established by a burst time plan which can be changed frequently.

TDMA Reference Burst   TDMA and SS/TDMA systems operate with a reference
burst which has the main function of establishing network timing; in all present
systems, reference bursts are uplinked from Reference Stations.   TDMA Reference
Functions are best implemented at the central node of the system, in the satellite;
there will be an on-board generator and modulator to transmit the reference burst
for each network. By dispensing with these specific uplinks, rain fades and
interference on the uplinks are eliminated; network AVAILABILITY is improved.
With due attention to control of this on-board subsystem of clocks (but with ground-
based atomic stabilities) and synchronization aids, such centralization can provide
SIMPLIFICATION in the provision of TDMA reference functions. It is also expected
that there would be ECONOMIES in earth station equipment and operations when
on-board reference signals are used.

Regeneration   Regeneration for TDMA is equivalent to burst-mode
demodulation of the 120-Mbit/s QPSK signal in P and Q channels; this is to be done
in the satellite for (at least) six pairs of signals, and each P and Q pair modulates a
downlink carrier for transmission in its designated beam. It is conceivable that only
some links would be regenerated, and switched through the MSM.

For the downlink, there is a benefit of up to 3 dB (aggregated). For the uplink,
there is a reduction of EIRP (saving of 10 dB, likely in the HPA); this leads to a
significantly LOWER COST OF THE EARTH SEGMENT. Such separate and
BETTER LINKS can have HIGHER QUALITY, or HIGHER AVAILABILITY, or a
chosen mix of both benefits. Alternatively, the design could be for MORE
CAPACITY of the whole system, or LESS SATELLITE POWER could be installed to
reduce the mass of the bus. Regeneration provides DESIGN FLEXIBILITY for links
in the network, and there is EASE OF OPERATION; with the SSPA/TWTA
operating point established on the satellite, there is little effect from uplink fades and
the downlink signal is uniform. This could be a major benefit in Ku bands, and
likely essential in Ka bands.

Multi-Carrier Demux, Demod   Regeneration for IDR signals in FDMA format consists of frequency demultiplexing many carriers in (a portion of) a transponder, and (bulk) multi-carrier demodulation (MCDD) to recover the symbol streams. It is necessary to demultiplex a variety of carriers which typically have a mix of information rates, from 64 kbit/s to 2.048 Mbit/s, and use coded QPSK modulation; applications have in a reasonably flexible frequency plan to accommodate changing service demands. Work is on-going to investigate methods to minimize power consumption in configurations with a fully flexible frequency plan, as well as for uniform channelization. The preferred technology uses CMOS digital signal processing, which handles large complex designs at low power consumption. Demultiplexing utilizes efficient digital filtering and Fast Fourier Transform algorithms. The demodulation approach explores time-shared hardware among the many IDR carriers, and methods to reduce the requirements for power-intensive interpolation filtering. As an admittedly aggressive goal, the complete MCDD with asynchronous clocks at several rates will be implemented on several ASIC[3] chips and is targeted for about 25 W of power for a 36-MHz bandwidth.

Uplink Decoding   FEC coding gains of several deciBels can be obtained by decoding in the satellite. Regenerative benefits are all present with MCDD and decoding, plus there is enhanced robustness to interference, since the uplink degradations can be taken out effectively; an example in Table 3 illustrates this. An uplink fade affects only that uplink, not the operating point of the satellite HPA, and uplink power control can be applied very effectively to maintain the uplink quality.

Time-Division Mux   When used in conjunction with MCDD and decoding, more significant link gains and flexibility result from using time-division multiplexing (TDM) in the downlink. If used for the entire transponder, almost no output backoff is needed and there are no intermodulation products, since there is only one carrier;

---

[3]   Application Specific Integrated Circuit, usually in complementary metal–oxide–silicon (CMOS) technology.

if not used for the whole bandwidth, TDM makes possible fewer downlink carriers, so that intermodulation can be controlled with minimal backoff. Compared to FDMA with transparent transponders, regeneration by MCDD possibly with decoding, followed by TDM downlink(s) provides many dBs of LINK IMPROVEMENTS. For example, when operating with a representative satellite (C band, 36 MHz, 36 dBW, hemi beams), MESH networks of IDR stations and VSATs, all using 1.8 m antennas, QPSK, rate-1/2 soft-decision FEC, are downlink power limited. Table 3 records some interesting results when interpreted as transponder capacity, indicated in numbers of 64 kbit/s channels with the indicated end-to-end BER for more than 99.96% availability (Tables 1 and 2). For the transparent case, the operating point was selected for maximum capacity; and for regenerative cases, the uplink was selected to be 3 dB better than the downlink. For all regenerative cases, the uplink power could be reduced by about 10 dB compared to the transparent case. Other link parameters were kept the same.

Networks using MCDD/TDM will incur costs not only for the payload but at the earth stations too: it is noted that receivers need to be equipped for the wider bands but comparable E_b/N_o, and the recovered TDM streams must be demultiplexed. It is recognized that compared to TDMA, an MCDD/TDM system does not require precise synchronization, which can be a strong advantage for economical small terminals.

Table 3: COMPARISONS of CAPACITIES (No. of 64-KBIT/S CHANNELS)
for TRANSPARENT and REGENERATIVE LINKS

Bit Error Ratio	Transparent FDMA / FDM	Regenerative FDMA, MCDD / FDM	Regenerative MCDD+Dec / Enc+FDM	Regenerative MCDD+Dec / TDM+Enc
1.0E-6	72	109	147	211
1.0E-3	123	161	246	354

Baseband Switch Matrix   Baseband switching is circuit routing done on sets of symbol streams; for example, P and Q streams can be switched in parallel matrices in the so-called "space" switch (like a cross-bar configuration). Very HIGH CONNECTIVITY can be achieved for systems of large size: N x N, where N can be as high as 30; a developmental model uses modules of size 16 x 4, which are combined to form 16 x 16 matrices. These modules are fabricated on GaAs substrate and use advanced active devices. As an MSM alternative, a BSM can work with regenerated TDMA streams. For MCDD/TDM systems, BSMs would provide the cross-transponder connectivity and cross-strapping between C and Ku bands (and Ka bands). All BSM applications could be designed for LOWER MASS OF THE PAYLOAD, compared to other switching methods.

Baseband Processing (BBP) Given regeneration (and possibly decoding), a BBP works on the symbol streams, usually in 8-bit increments, to provide a very high SERVICE FLEXIBILITY; in effect a BBP is a "time-stage" switch since incoming signals are buffered and then read out according to the desired plan in the downlink. It is possible to selectively broadcast or "narrowcast" a signal or a portion of it, where the target stations are in different beams and bands. Likewise, a baseband processor can selectively multiplex two or more uplink signals for eventual reception at a single or many sites (e.g. data gathering, file sharing). With the BBP already developed and tested by INTELSAT, it is possible to interconnect low-rate TDMA and thin-route FDMA traffic which includes the necessary RATE CHANGES. The above flexibility extends over time, since the BBP could be reprogrammed (uploaded from ground) for revised and new services. Also, for a given connectivity requirement, FEWER UPLINKS can be anticipated since multi-destinational signals are uplinked only once; this results in system and economic advantages.

Technology Readiness   These advanced technologies evidently have varying degrees of readiness for space communications use. Burst-mode modulators and related equipment for TDMA Reference Bursts are fully developed, tested, but it is prudent to prepare engineering models to verify compatibility and interoperability with existing networks: these can be ready by end 1994. Similarly for regenerators for wideband signals, (burst or continuous), engineering models already exist or can be finished in a few years. Advanced forms of MSMs and BSMs are considered mature for space use; but it is appropriate to note that production yields of BSM chips need to be improved and verified. MCDDs can be in two forms: those that are ready today are suitable for fixed, uniform FDMA formats and synchronous signals; those with more flexibility and for asynchronous signals are in development. Flight-readiness of the regenerative MCDD/TDM payloads is being assessed, with space qualified technology to be available in 1995. Uplink decoders are of the same processing types and are also expected to be mature by 1995. The more advanced BBP will likely need further development to representative engineering models before carriers and contractors will commit to them.

6. ADVANCED METHODS FOR SATELLITE BUS AND ORBITS

There are at least two topics not usually discussed along with telecommunications systems, but which will have very significant impact on payloads, services and economics of the future.

Electric Propulsion Systems

The satellite industry has long had an interest in electric propulsion (EP), and INTELSAT regards this as a potential technology for satellites starting their operational lives over the next decade. Through the use of EP systems, the potential for MASS SAVINGS is greater by far than that of any other technology. Lower mass can be translated into decreased launch cost, or increased payload, or longer maneuver life, as desired.

Electric propulsion devices can be grouped broadly into three classes, depending on the mechanism for applying electrical energy: electrothermal, electromagnetic, and electrostatic. Devices from each class have either already been applied to spacecraft propulsion or are being prepared for use in the near term. Electromagnetic propulsion, in the form of the teflon pulsed plasma thruster, was used for east-west stationkeeping on spinning satellites (LES series) in the 1960s. The simplest form of electrothermal propulsion, the hydrazine resistance jet, has been operational for over five years; the chemical energy of the monopropellant is enhanced by ohmic heating, with a resulting increase in specific impulse I_{sp} from 220 to 300 seconds. The more sophisticated hydrazine arcjet which generates I_{sp} in the 500 - 550 sec range, will be operational on the Telstar satellites. The so-called SPT (Stationary Plasma Thruster) developed in the Soviet Union, and flown on a number of short duration military satellites, is a version of a gridless electrostatic ion thruster; typical I_{sp} is 1600 - 1700 sec. Programs are presently underway to demonstrate adequate SPT lifetime, to upgrade the power conditioners, and to define the thruster plume so as to minimize its interaction with adjacent spacecraft components [14]. At the higher end of the I_{sp} range (2500 - 4000 sec) is the classical gridded electrostatic ion thruster, several versions of which are being developed for north-south stationkeeping (NSSK) on satellites which are scheduled for launches in the mid 1990s. Germany has developed its version of the ion thruster, which uses RF excitation to produce ionization, for use on both the EUREKA and ARTEMIS satellites. In Japan, a modified version of the classical Kaufman type electron bombardment thruster is being readied for the ETS-6 satellite. The United Kingdom is preparing its ion thruster version, also a modified Kaufman type, for use on ARTEMIS. In the USA, an advanced form of the electron bombardment ion thruster, the so-called ring-cusp design which uses a "picket-fence" magnetic field rather than a solenoidal field, is being life tested.

Although electric propulsion can be used for any low-thrust satellite maneuver, the most significant mass benefits result from application to the maneuvers entailing the highest velocity increments: NSSK (50 m/s per year) and orbit raising (500 - 5000 m/s). In the case of NSSK, it is generally accepted that the electric propulsion devices will be powered by the batteries already on board rather than by dedicated solar arrays; hence, only the masses of the propulsion and power conversion subsystems need to be considered. For NSSK, small satellites and short lifetimes favor the lighter, lower I_{sp} devices, while large satellites with plenty of power and long lifetimes would favor ion propulsion. In the case of orbit raising, the EP devices must be powered by solar arrays. In the foreseeable future, high-altitude orbit raising, using only the power and propulsion already on board, could be applied; here, a short mission time is just as important as lower mass, and the lower I_{sp} devices which generate higher thrust with a given amount of power are favored. When a full LEO/GEO maneuver is seriously entertained (approximately 6000 m/s), the higher benefits of the ion thrusters will probably be the dominant factor.

Collocation of Satellites

Another potential approach for future systems is collocation of GEO satellites; pairs of present-size satellites would occupy existing orbital slots. Since available orbital locations are limited, increased capacity per orbital location is highly desirable. To achieve similar capacity using a single large spacecraft could involve major satellite bus development effort and long delivery lead time. Also, satellite collocation allows for a flexible deployment strategy of smaller and staggered launches, for faster augmentation of services at locations where traffic demand has increased.

In a feasibility study, we concluded that no major problems are anticipated for collocation scenarios with satellite separations of 0.03° or more. However, if it is desired to enable a single large earth station antenna to access the pair simultaneously (a strong economics argument), the satellites must be situated closer. Such close "binary" satellites will need new control strategies. Relative disturbances from NSSK maneuvers for instance must be detected and corrected almost immediately, before the relative separation and drifts have a chance to grow out-of-bounds. These disturbances, for example thruster uncertainties, must be controlled where possible. In addition, better equipment, for instance for accurate satellite position determination, must be in place.

The orbital dynamics and operational aspects are being addressed. The challenge is to identify control strategies which minimize the impact on present operations. The type of services, frequency bands, and traffic connectivity issues at a particular orbital location will strongly impact the type of satellite collocation scenarios appropriate for that location, and also in turn the type of processing payload required.

7.   CONCLUSION

INTELSAT is making rapid progress as part of the general trend to provide digital services. Our emphasis on **quality** and **availability** is in evidence as we prepare for more stringent standards and competitive markets. The key foundations of GEO satellite systems are COVERAGE and CONNECTIVITY, and advanced technologies will enable both to be enhanced.

Through use if trellis-coded modulation, there are possibilities for better SPECTRAL EFFICIENCY and more CAPACITY using the larger earth stations. Intelnet, IBS and VSAT services can be enhanced using advanced technologies to provide practical end-to-end services using customer-premise earth stations. Such networks will be further improved through new protocols, including on-demand capacity, whether in FDMA or TDMA.

As a brief conclusion on OBP from a service perspective, we note the advantages from SS/TDMA, reference burst and TDMA regeneration: LINK IMPROVEMENTS, CONNECTIVITY, and earth station ECONOMIES. Of course, IDR, IBS and VSAT services will benefit immensely from regeneration, decoding, switching and fewer downlinks: PERFORMANCE, CAPACITY and EFFICIENCY. At higher rates, digital TV services could get commensurate advantages from similar processing. Looking to prospective services, MCDD/TDM technologies are very compatible with digital audio broadcasting, for example; and BBP has the FLEXIBILITY to provide circuit-switched and possibly packet-switched functions in emerging networks.

It is thus possible to envision several kinds of "power" in orbit: we anticipate more (linear) power in future HPAs; larger and more flexible switches will be in payloads; and electric propulsion is an exciting new form of real power. Major advances in processing of traffic signals are yet another form of power in future INTELSAT satellites.

Acknowledgments

The author gratefully acknowledges the support from INTELSAT Management and the specific expert contributions and comments from colleagues: R. Schweikert (INTELSAT Assignee) on TCM and outer codes, T. Abdel-Nabi on service requirements, B. Free (consultant) on electric propulsion, and P. Yeung on collocation.

References

1   John D. Hampton, "Implementing the Digital Decade", Keynote Address, Proceedings, ICDSC-9, Copenhagen, May 1992.

2   G. Forcina, W.S. Oei, T. Oishi and J.F. Phiel, "INTELSAT Digital Circuit Multiplication Equipment", Proceedings, ICDSC-8, Guadeloupe, April 1989.

3   G. Forcina and T. Oishi, "Facsimile Compression in the INTELSAT DCME", Proceedings, ICDSC-9, Copenhagen, May 1992.

4   CCITT, (draft) Recommendation G.728, COM-XV-R-G7-E, "Coding of Speech at 16 kbit/s using Low-Delay Code-Excited Linear Prediction (LD-CELP)", December 1991.

5   W.S. Oei and S.P. Tamboli, "International Fixed Satellite Systems in Synchronous Digital Hierarchy Transport Networks", Proceedings, ICDSC-9, Copenhagen, May 1992.

6   T. Oishi and D. Gokhale, "High-Speed Satellite Efficient X.75 Protocol Converter", Proceedings, ICDSC-9, Copenhagen, May 1992.

7   SDM-140 Satellite Modem Specification, EFData Corporation, Arizona, USA, 1991.

8   P. Nuspl, R. Peters, T. Abdel-Nabi and N. Mathews, "On-board Processing for Communications Satellites: Systems and Benefits", INTERNATIONAL JOURNAL OF SATELLITE COMMUNICATIONS, Vol.5, No.2, pp.65-76, Wiley-Interscience, April-June 1987.

9   P. de Santis, "INTELSAT Research and Development in On-Board Processing Technologies", International Symposium, Satellite Communications: Present and Future, Odessa, USSR, October 1990.

10   P.P. Nuspl, G. Dong and H.C. Seran, "Laboratory Measurements of On-Board Subsystems", (Poster Paper), Proceedings, Second NASA Space Communications Technology Conference, Onboard Switching and Processing, Cleveland, pp.113-133, November 1991.

11   P.P. Nuspl and G. Dong, "On-Board Processing for Telecommunications Satellites", Proceedings, Second NASA Space Communications Technology Conference, Onboard Switching and Processing, Cleveland, pp.223-238, November 1991.

12   K. Betaharon, K. Kinuhata, P. Nuspl, and R. Peters, "On-board Processing for
     Communications Satellite Systems: Technologies and Implementations",
     INTERNATIONAL JOURNAL OF SATELLITE COMMUNICATIONS, Vol.5,
     No.2, pp.139-146, Wiley-Interscience, April-June 1987.

13   P.V. de Santis and P.S. Yeung, "On-Board Demultiplexing of Unrestricted
     FDMA Traffic", Proceedings, ICDSC-9, Copenhagen, May 1992.

14   Proceedings, INTELSAT Electric Propulsion Symposium, Sacramento, CA June
     1991.

ICDSC   International Conference on Digital Satellite Communications,
        co-sponsored by INTELSAT and the host Signatory.

PERSONAL HANDHELD COMMUNICATIONS VIA HYBRID Ka- AND L/S-BAND SATELLITES*

Russell J. F. Fang

COMSAT Laboratories

22300 Comsat Drive

Clarksburg, MD, 20871, U.S.A.

ABSTRACT

A system concept is described for personal communications between handheld units and public-switched telephone networks (PSTNs) via geosynchronous satellites. The handheld-to-satellite links are provided for land-based users at 20/30 GHz Ka-bands nominally under clear sky and shadow-free conditions and augmented by L- and S-bands under rainy or multipath fade conditions. These dual-mode handheld units are used in regions where heavy traffic is envisaged. For areas where low usage is expected, only L- and S-band handheld units will be utilized.

Handheld units employ patch antennas at Ka-band and stub antennas at L/S-band to provide an antenna gain of 3 dBi. Transmitter power is less than 0.6 W at L-band and 0.8 W at Ka-band. Frequency-division multiple access/code-division multiple access (FDMA/CDMA) is used in conjunction with binary phase shift keying (BPSK) modulation and Viterbi and Reed-Solomon forward error correction (FEC) coding. An adaptive code-excited linear prediction (CELP)-based codec at 2.4 kbit/s is employed for low-rate encoding of speech.

Spacecraft antennas of 5 and 9 meters are used for Ka- and L/S-bands, respectively, for communications into handheld units. These antennas generate cellular footprints on the land masses. A 0.5-meter Ku-band antenna is used for communications into gateway stations at 14 and 11.7 GHz. Satellites are positioned over land masses to achieve better than 20° elevation angle for most areas and hence to minimize potential shadowing problems. The availability of a wideband feeder link at Ku-band and all gateway stations in the same footprint of a particular satellite result in several benefits:

   (a) bent-pipe transponders can be easily interconnected with the cellular footprints at Ka- and L/S-bands;

   (b) the number of gateway stations required can be minimized;

   (c) the feeder link subsystem costs can be reduced.

The nominal capacity of such a satellite is about 30,000 full duplex channels.

CDMA is used to minimize antenna beam isolation requirements, to realize simultaneous demand assignment of satellite bandwidth and power, to achieve natural statistical multiplexing of voice and data at the transponder input, and to minimize transmission power spectral density and hence adjacent satellite interference and intersystem interference. Frequency-division multiple access is used to facilitate CDMA operation over the bent-pipe satellite transponders and to reduce CDMA chip rates, thereby minimizing DC power requriements for the handheld units and the size and weight of

---

* This paper is based upon work performed under the sponsorship of COMSAT Corporation. Views expressed are solely those of the authors and not necessarily those of COMSAT.

their power supplies. Traffic handoff between cellular footprints is employed to minimize satellite antenna pointing and hence stationkeeping requirements.

INTRODUCTION

Analog cellular mobile communications and cordless phones have experienced unexpected and explosive growth during the past decade. Their popularity and usage have increased tremendously, while the equipment price has dropped drastically. During the same period, analog and digital very large-scale integration (VLSI) technologies have greatly advanced so that lightweight and pocket-size phones have become a fad. Features such as display, redial, memory storage, call timer, status monitoring, voice dialing, hands-free operation, data/fax, and integration with a pager are all in demand by users. Services such as call forwarding, call transferring, conference call, voice mail, call screening, and encryption are becoming readily available. Ubiquitous handheld personal communication is clearly the trend of the future.

Due to the mobility of mobile and pocket-size handheld units, roaming agreements, roaming validation, intersystem call hand-off, and billing have become major issues for intersystem operations. In fact, these have been some of the major concerns in the design of the second generation digital cellular systems such as Pan-European Groupe Spatiale Mobile (GSM) [1] and North American EIA/TIA IS-54 [2] standards. However, terrestrial cellular and wireless communications are ideally suited only for serving populated areas. They are not economical for serving less populated remote regions. Instead, satellites can be the ideal medium for providing thin-route personal communications to these remote regions.

With its success in the cellular equipment business and its experience with satellite on-board baseband processor development gained through the NASA Advanced Communications Technology Satellite (ACTS) program [3], Motorola has recently proposed the well-publicized orbiting cellular system called IRIDIUM [4]. Worldwide, direct handheld-to-handheld communication is made possible by using 77 interconnected satellites in low earth orbits (LEOs) (with 11 satellites in 7 polar planes each), which form the moving cells in the sky. Following this proposal, other LEO satellite systems such as Globalstar [5] and Odyssey [6] were also proposed by Loral/Qualcomm and TRW, respectively.

To prove that handheld communications via geosynchronous earth orbit (GEO) satellites is also possible, Hughes [7] proposed a system which employed a 30-m L-band antenna on-board the spacecraft. Many other systems were also proposed for personal handheld communications including Project 21 from Inmarsat at L- band and PASS [8] from NASA Jet Propulsion Laboratory at Ka-band.

Designing a satellite with a 30-m antenna structure obviously poses a major technical challenge in terms of both fabrication and deployment. A personal communications system such as PASS at Ka- band requires a significant amount of margin to compensate for rain fades and other multipath effects.

In this paper, a hybrid L/S- and Ka-band GEO satellite system concept is proposed for personal communications between handheld units and public-switched telephone networks (PSTNs) through Ku- band gateway stations. Potential applications include telephony, group-3 facimile, radio dispatch, paging, data broadcasting, data collection, and message delivery.

SYSTEMS CONCEPT

The basic system concept is as depicted in Figure 1. Single-mode handheld units at L/S-band, or dual-mode handheld units at L/S- and Ka-bands, communicate over geostationary satellites into the PSTNs through Ku-band gateway stations. Submarine cables or other terrestrial links are used to extend the accessibility of gateway stations to handheld units on the other continents.

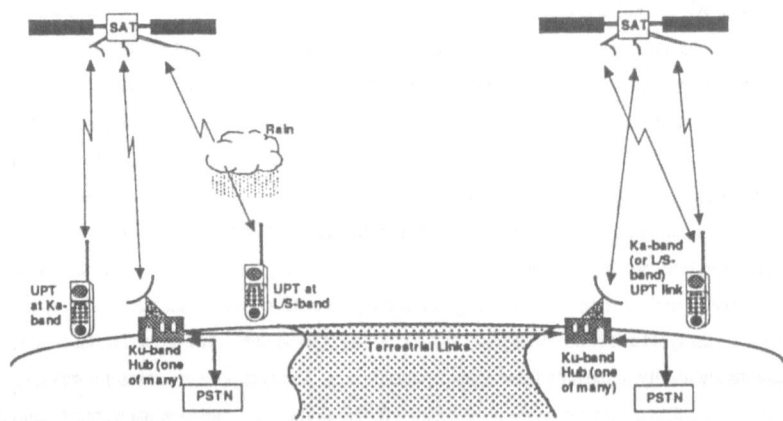

Figure 1. Hybrid Satellite System Concept

Satellites are located over land masses in order to minimize footprint territorial coverages for a given spacecraft antenna size and hence to increase the system capacity. By locating satellites over land masses, the scan loss of the spacecraft antenna can also be minimized when only heavy traffic areas need to be covered as indicated in Figure 2. Due to ample bandwidth (1 GHz) available at Ka-band, narrow spot beams are used to cover these regions so that a large

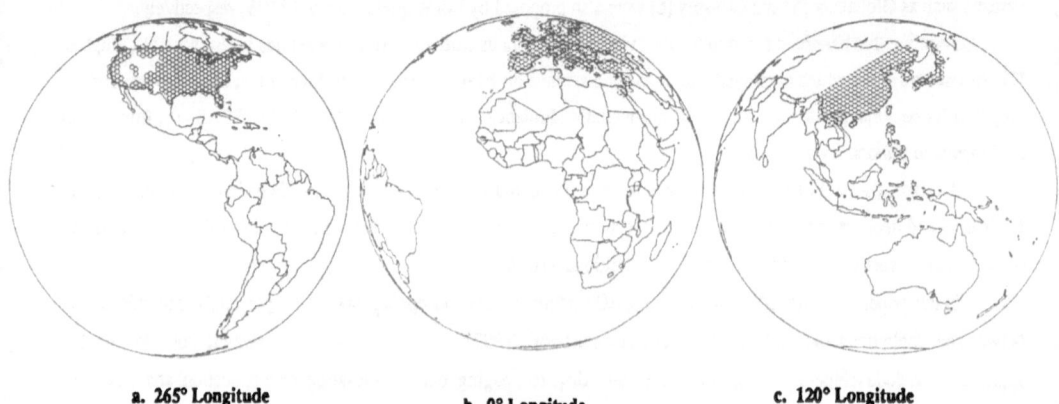

a. 265° Longitude   b. 0° Longitude   c. 120° Longitude

Figure 2. Ka-Band Footprints

amount of traffic can be supported during clear sky and clear path conditions. Moreover, smaller RF electronics may be used at Ka- band to reduce the size and weight of the handheld units. In fact, in the future, a more directive array antenna can be developed on a small patch to increase overall link margin, system capacity, and data rates.

Narrow L- and S-band antenna beams are used to cover all land masses as shown in Figure 3. These beams augment Ka-band coverages, take over handheld traffic in Ka-band coverage areas during rain fades or multipath shadowing conditions, and carry overflow traffic from Ka-band coverage areas. (Broad L-band beams are available from the current Inmarsat satellites to cover ocean regions for serving large mobile platforms such as Inmarsat-B, -M, or Aero terminals.) Since satellites are positioned over land masses, the elevation angle for most of the land mass areas is better than 20°, as can be seen from Figure 4. Therefore, the troublesome shadowing problem can be significantly reduced for land-based handheld communications.

Initial satellite acquisition and signaling, as well as call processing at handheld units, is accomplished at L/S-band. The overall link availability is totally determined by the L/S-band link, even for those users who are also served by Ka-band, because Ka- band is used only under clear sky or clear path conditions.

L-band up and S-band down are chosen for the handheld- to- satellite communications mainly to achieve three objectives:

(a) to avoid the potential passive intermodulation problem facing all 1.6/1.5-GHz L-band satellite antenna designers

(b) to avoid the potential costly diplexer design for a low-cost 1.6/1.5-GHz L- band handheld unit

(c) to avoid the use of a time-division duplexing (TDD) scheme such as that used in the IRIDIUM system to solve the passive intermodulation problem, since a 3 dB higher C/N_0 will be required as a result of the factor of 2 increase in information rate through TDD.

Ku-band is selected for feeder links mainly because of spectral availability. By employing linear polarization reuse, a total 1-GHz spectrum can be made available in four 250-MHz chunks at 14 and 11.7 GHz, sharing with the fixed system operators. With this amount of spectrum available, simple global beams can be used to cover all Ku-band gateway

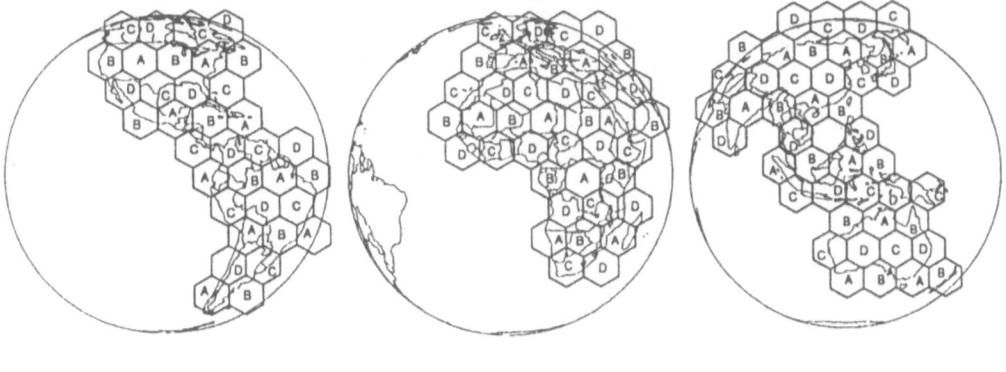

a. 265° Longitude   b. 0° Longitude   c. 120° Longitude

Figure 3. L- and S-Band Footprints

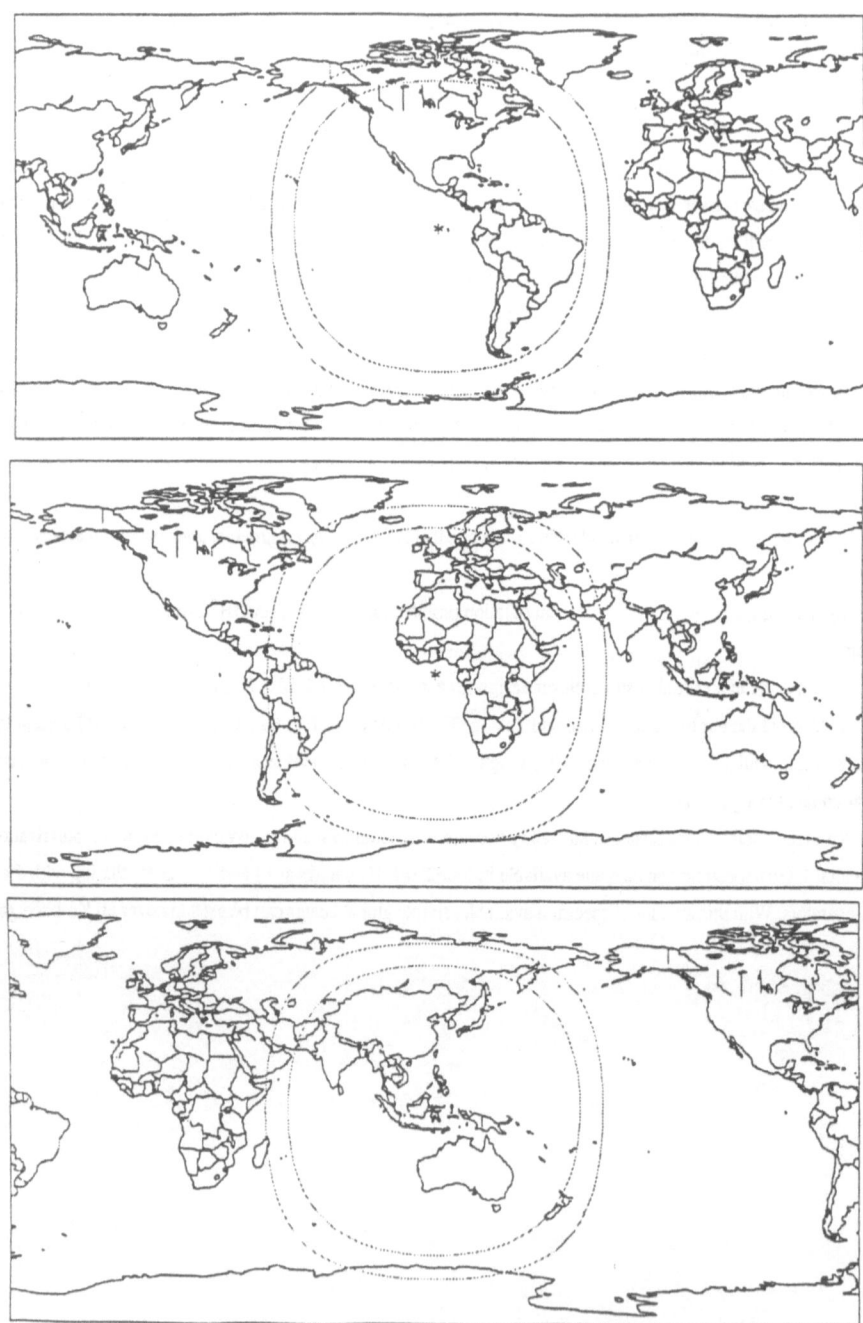

Figure 4. 10° and 20° Elevation Contours for the Three Satellite Views

stations, as illustrated in Figure 5. This global beam coverage also mitigates the potential satellite on-board beam inter-connect problem inherent in all multibeam satellite systems.

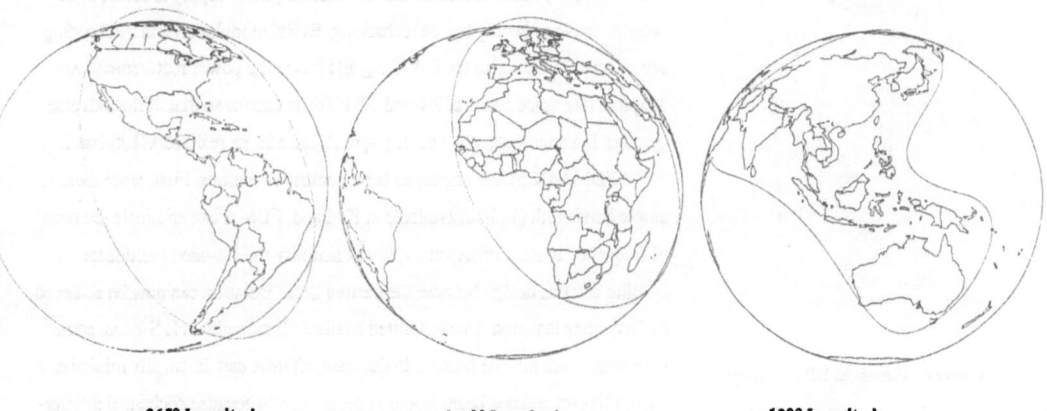

| a. 265° Longitude | b. 0° Longitude | c. 120° Longitude |

Figure 5. Ku-band Shaped "Global" Coverages

HANDHELD UNITS

The main characteristics of the handheld unit, which is artistically illustrated in Figure 6, are listed below (assuming 1995 technologies):

Frequency:
- •Ka-band: 30 GHz transmit; 20 GHz receive
- •L/S-band: 1.6 GHz transmit; 2.6 GHz receive

Polarization:
- •30 GHz: left-hand circular polarization (LHCP)
- •20 GHz: right-hand circular polarization (RHCP)
- •1.6 GHz: LHCP
- •2.6 GHz: RHCP

Antenna:
- •Omnidirectional: 3-dBi gain
- •Ka-band: patch
- •L/S-band: stub

G/T sensitivity:
- •Ka-band: -19.3 dB/K
- •S-band: -15.8 dB/K

HPA power:
- •Ka-band: 0.8
- •L-band: 0.6 W

Access method: FDMA/CDMA

Modulation: BPSK

FEC coding: Rate-1/2 convolutional innercode with Viterbi decoding and Reed-Solomon (255, 223) block outer code.

E_b/N_0 @ BER = 10^{-2} : 2.0 dB

Voice codec: 2.4-kbit/s CELP with 1.2-kbit/s capability under degraded link conditions.

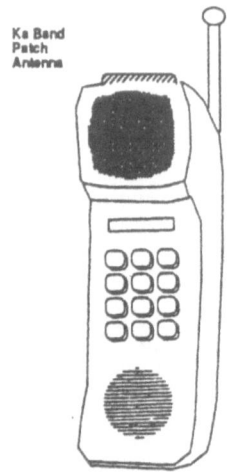

Ka Band
Patch
Antenna

L/S Band
Stub
Antenna

Figure 6. Handheld UPT Concept

Omnidirectional antennas and circular polarization are employed to eliminate any antenna pointing requirements and minimize the size, weight, and cost of the handheld units. Less than 1-W high-power amplifier (HPA) average power is used to ensure that the handheld unit is well within the radiation safety limit for human use and that the power supply is small, light-weight, and long lasting before recharging. BPSK modulation and FEC coding are selected to minimize the link C/N_0 and hence the power requirement. An adaptive rate voice codec at 2.4 and 1.2 kbit/s is chosen so that during adverse propagation conditions good quality speech can still be realized at L/S-band.

FDMA/CDMA is employed for a number of reasons. First, since there is ample bandwidth (1 GHz) available at Ka-band, FDMA can eliminate the need of frequency reuse. Consequently, it will simplify the Ka-band multibeam satellite antenna design because the desired beam isolation can now be achieved by frequency isolation. Due to limited available bandwidth at L/S-band, some frequency reuse may be needed. In this case, CDMA can drastically minimize the multibeam antenna beam isolation requirements because cochannel interfering signals from the adjacent beams will be drastically reduced after the despreading operation at the CDMA receiver. Therefore, FDMA/CDMA can significantly simplify the satellite multibeam antenna design by eliminating the difficult beam isolation requirements and consequently antenna efficiency can be greatly improved.

Secondly, voice-activated CDMA can facilitate simultaneous demand assignment of both satellite transponder bandwidth and power—unlike voice-activated single-channel-per-carrier (SCPC) systems where only transponder power (but not bandwidth) is shared on demand by the active users, and also unlike voice- activated time-division multiple-access (TDMA) systems where only bandwidth is shared, but not power. Thirdly, CDMA voice and data traffic are statistically multiplexed together at the satellite transponder in a natural manner, thereby automatically realizing the best possible activity compression gain for any mixture of voice and data. Finally, CDMA can lower the transmission power spectral density, thereby significantly reducing the interference into adjacent satellites or into other terrestrial systems.

SPACECRAFT CONCEPT

An artist's concept of the spacecraft is sketched in Figure 7. It is envisioned to be a 3-axis-stabilized satellite with two large antenna reflectors serving handheld units, one 9-m mesh for L- and S-band and the other 5-m dish for Ka-band. The feeder link gateway stations are served by a 0.5-m Ku-band antenna. Key spacecraft parameters are given below.

Frequency:
- •Spacecraft-to-handheld links:
  - —Ka-band: Transmit @ 20 GHz (RHCP)
    - Receive @ 30 GHz (LHCP)
  - —S-band: Transmit @ 2.6 GHz (RHCP)

Frequency (cont'd):
- —L-band: Receive @ 1.6 GHz (LHCP)
- •Spacecraft-to-gateway stations links:
  - —Ku-band: Transmit @ 11.7 GHz (VP & HP)
    - Receive @ 14 GHz (VP & HP)

Antenna:

- •Spacecraft-to-handheld links:
- —Ka-band transmit/receive: 5-m dia.
- —S/L- band transmit/receive: 9-m dia.
- •Spacecraft-to-gateway station links:
- —Ku-band transmit/receive: 0.5-m dia.

G/T sensitivity:

- •30 GHz: 29.2 dB/K

G/T sensitivity (cont'd):

- •1.6 GHz: 9.6 dB/K
- •14 GHz: -6.5 dB/K

Transponder e.i.r.p.:

- •20 GHz: 61 dBW
- •2.6 GHz: 49 dBW
- •11.7 GHz: 34 dBW

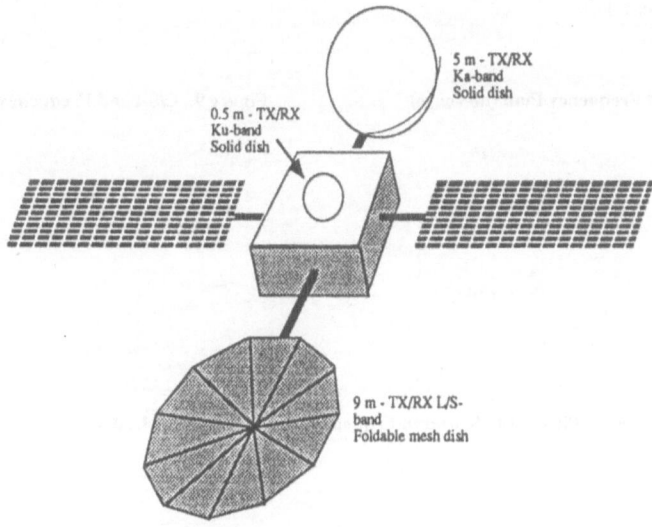

5 m - TX/RX
Ka-band
Solid dish

0.5 m - TX/RX
Ku-band
Solid dish

9 m - TX/RX L/S-
band
Foldable mesh dish

Figure 7. Spacecraft Concept

The sample frequency plans are illustrated in Figures 8–10 for Ka-, L/S-, and Ku-bands, respectively. The bent-pipe transponder architectures are shown in Figures 11 and 12 for the handheld forward and return links, respectively. S-band matrix amplifiers are employed to share satellite DC power between down-link beams. The multiplexer filter bank can be implemented by surface acoustic wave (SAW), chirp-transform, or digital signal processing technologies depending upon the final design tradeoffs.

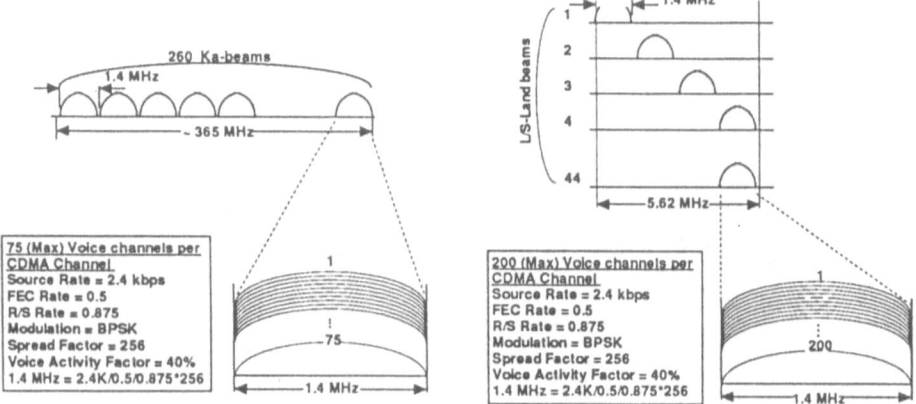

Figure 8. Ka-band Frequency Plan (no reuse)   **Figure 9. L/S-band Frequency Plan**

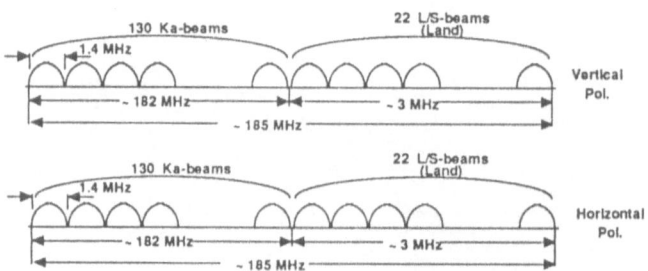

Figure 10. Ku-band Fequency Plan for Gateway Links

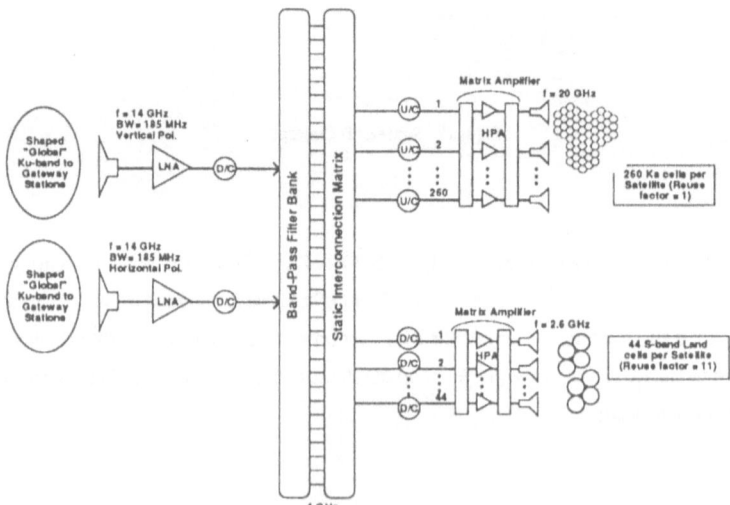

Figure 11. Forward Link Architecture

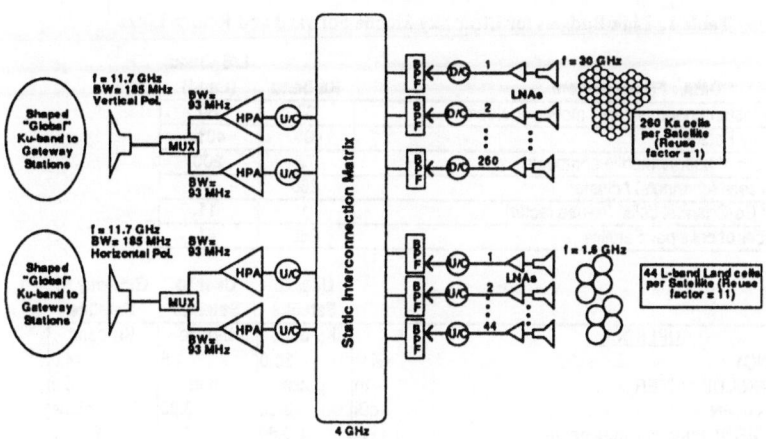

Figure 12. Return Link Architecture

LINK BUDGETS, CAPACITY, AND MASS & POWER ESTIMATES

Link budgets for the forward and return links are summarized in Table 1. The L-band handheld up-link is significantly stronger than the Ka-band handheld up-link in terms of C/N_0, mainly due to the use of a larger satellite antenna.
The feeder link gateway stations are assumed to use 5.5-m antennas. For those locations with a significant amount of rain
precipitation, 7-m and/or diversity antennas may be used as deemed necessary.

The nominal system capacity per satellite is estimated at about 30,000 duplex channels with about 20,000 provided by Ka-band.

A rough estimate of the satellite mass and power budget is also given in Table 2 by assuming 1995 technologies.
It appears that a Titan 3 class launcher would be needed.

CONCLUDING REMARKS

For satellites with very small footprint (0.3°) coverages, tolerance to satellite antenna pointing error for
stationkeeping is extremely small and can be very difficult to realize in the conventional geostationary satellite design.
Since traffic handoff is a must for personal handheld or mobile communications, it can be employed to drastically relax
the satellite pointing accuracy requirements, as long as the handheld units are covered by some other beams even though
the satellite antenna pointing may be off by a beamwidth or more. Note that the daily periodic movement of the
spacecraft in the geosynchronous orbit is extremely slow and limited, except during maneuvering. (The satellite is
practically stationary during any given telephone call.) However, movement during maneuvering can be minimized by
employing "critical damping" in the design. With this observation, satellite stationkeeping can be made much simpler and
also consume significantly less fuel.

Table 1. Link Budgets for Clear Sky Mobile Forward and Return Links

CDMA / FDMA system	Ka-band	L/S -band (Land)
Number of Instantaneous users / mobile spotbeam	30	80
Voice Activity Factor	40%	40%
Total Number of users / mobile spotbeam	75	200
Number of cells (channels) / cluster	260	4
Number of Co-Channel cells (re-use factor)	1	11
Total number of cells per Satellite	260	44

UPLINKS		User to Satellite	User to Satellite	Gateway to Satellite
		Ka-band	L-band	Ku-band
FREQUENCY	GHz	30.0	1.6	14.0
ES ANTENNA DIAMETER	m	patch	stub	5.5
ANTENNA GAIN	dBi	3.00	3.00	55.54
HPA POWER at antenna flange/beam	W	0.80	0.60	279.25
EIRP / beam	dBW	16.80	19.81	80.00
PATH LOSS (at 10°)	dB	214.66	188.90	207.84
Satellite Antenna Diameter	m		9.0	0.5
Satellite 3dB beamwidth	deg.	0.30	1.87	3.00
Satellite Antenna Gain (at edge)	dBi	56.30	35.00	20.25
SATELLITE G/T	dBi/K	28.10	8.70	-6.47
BOLTZMAN CONSTANT	dBW/Hz-K	-228.60	-228.60	-228.60
UPLINK C/No /beam	dB - Hz	58.84	68.21	94.29
UPLINK C/No / CDMA channel	dB - Hz	58.84	68.21	69.46
UPLINK C/No / USER	dB - Hz	58.84	68.21	-

DOWNLINKS		Satellite to Gateway (From Ka)	Satellite to Gateway (From L)	Satellite to User	Satellite to User
		Ku-band	Ku-band	Ka-band	S-band
FREQUENCY	GHz	11.7	11.7	20.0	2.6
Satellite Antenna Diameter	m	0.5	0.5	5.0	9.0
Satellite 3dB beamwidth	deg.	3.59	3.59	0.30	1.90
Satellite Antenna Gain (at edge)	dBi	18.69	18.69	51.80	39.20
HPA POWER at antenna flange /beam	W	31.60	31.60	7.80	7.80
EIRP / CDMA channel	dBW	12.40	20.10	60.00	47.40
PATH LOSS (at 10°)	dB	206.00	206.00	210.60	192.90
BOLTZMAN CONSTANT	dBW/Hz-K	-228.60	-228.60	-228.60	-228.60
RECEIVE ANTENNA DIAMETER	m	5.5	5.5	patch	stub
RECEIVE ANTENNA GAIN	dBi	53.98	53.98	3.00	3.00
RECEIVER G/T	dB/K	30.93	30.93	-19.29	-15.76
DOWNLINK C/No / CDMA channel	dB - Hz	65.93	73.63	58.71	67.34
TOTAL C/No / CDMA channel	dB - Hz	58.06	67.11	58.36	65.26
TOTAL C/No /USER	dB - Hz	43.29	48.08	43.59	46.23
TOTAL AVAILABLE C/No+Io /USER	dB - Hz	41.71	41.44	41.91	40.97
CARRIER INFORMATION RATE	KBits/s	2.4	2.4	2.4	2.4
OVERALL FEC CODING RATE		0.437	0.437	0.437	0.437
BPSK TRANSMISSION RATE	Ksps	5.489	5.489	5.489	5.489
AVAILABLE Eb/No /USER	dB	7.91	7.64	8.11	7.16
REQUIRED Eb/No (@BER=10-4)	dB	2.0	2.0	2.0	2.0
SYSTEM MARGIN	dB	5.906	5.638	6.108	5.165
CHANNELIZATION BANDWIDTH (40% excess)	KHz	7.684	7.684	7.684	7.684
DS-CDMA Spread Factor		256.0	256.0	256.0	256.0
PN SPREAD SPECTRUM BANDWIDTH	MHz	1.405	1.405	1.405	1.405

Table 2. Mass and Power Estimate for Satellite

	Mass (Kg)	Power (W)
Ka-band		
Antenna and Feed Network	75.00	
Amplifiers (HPAs)	118.18	7426.47
Receivers (LNAs)	118.18	780.00
L/S-bands		
Antenna and Feed Network	64.23	
Amplifiers (HPAs)	101.20	526.10
Receivers (LNAs)	66.00	132.00
Ku-band (feeder link)		
Antenna and Feed Network	1.60	
Amplifiers (HPAs)	2.00	90.34
Receivers (LNAs)	1.50	3.00
Other Subsystems		
Signal Processing (Filters, Converter	25.00	100.00
TT & C		40.00
Attitude control		30.00
Propulsion and Thermal		40.00
Power and Harness		25.00

	Mass (Kg)	Power (W)
Total for Communications Payload	572.89	9192.91
Prime Power Mass (Kg)	735.43	
Payload Allocation (Kg)	1308.32	

Using Late 90's Technology

Dry Mass (Kg)	2180.54
Propellant mass (Kg)	231.41
BOL Mass (Kg)	2411.95
Transfer Orbit Mass (Kg)	4389.75
Largest allowable TOM (Kg)	5700.00
Design Margin of (Kg)	1310.25

It should be noted that handheld users can avoid the shadowing problem when making calls to the PSTN users. They can also avoid the shadowing problem when receiving calls from PSTN users by judiciously applying paging. When a handheld user is on a mobile platform, a more directive antenna on the platform can be made available to significantly alleviate the problem.

ACKNOWLEDGMENTS

The author would like to thank Messrs. David Haschart and William Sandrin for their invaluable contributions to this paper.

REFERENCES

[1] V. Lev and M. D. Kotzin, "The Pan-European Digital Cellular (GSM) System," NASA/JPL Workshop on Advanced Network and Technology Concepts for Mobile, Micro, and Personal Communications, NASA/JPL, Pasadena, California, USA, May 1990, Proc., pp. 209-218.

[2] Electronic Industries Association, "Dual-Mode Mobile Station-Base Station Compatibility Specification," EIA/TIA Intrium Standard 54, December 1989.

[3] F. M. Naderi and S. J. Campanella, "NASA Advanced Communications Technology Satellite (ACTS): An Overview of the Satellite, the Network, and the Underlying Technologies," 12th AIAA International Communications Satellite Systems Conference, March 13-17 1988, Proc., pp. 204-224.

[4] R. J. Leopold, "Low-Earth Orbit Global Cellular Communications Network," Mobile Satellite Communications Conference, Adelaide, Australia, August 1990, Proc..

[5] R. K. Kwan and R. A. Wiederman, "Globalstar: Linking the world Via Mobile Connections," IEE International Symposium on Personal, Indoor, and Mobile Radio Communications, London, England, September 1991, Proc., pp. 318-323.

[6] R. J. Rusch, P. Cress, M. Horstein, R. Huang, and E. Wiswell, "Odyssey, A Constellation for Personal Communications", 14th AIAA International Communication Satellite Systems Conference, March 22-26, 1992, Washington, DC, USA.

[7] Hughes Space and Communications Group, "TRITIUM- An Advanced Mobile Satellite System," US FCC Filing, June 3, 1991.

[8] P. Estabrook et al., "A 20/30-GHz Personal Access Satellite System Design," IEEE International Conference on Communications, Boston, Massachusetts, June 1989, Proc., pp. 7.4.1-7.4.5.

COHERENT OPTICAL SPACE COMMUNICATIONS

Walter R. Leeb
Institut für Nachrichtentechnik und Hochfrequenztechnik
Technische Universität Wien
Gusshausstrasse 25, A-1040 Wien, Austria

ABSTRACT

This paper reviews the present state of concepts and technologies of coherent optical space communications. Emphasis is placed on the receiving segment. Heterodyne and homodyne reception allows to realize sensitivities close to the theoretical limits which amount to as little as 10 to 40 photons per bit. They ask for a narrow-linewidth laser not only at the transmitter terminal but also as local optical source within the receiver. Coherent receivers require tight frequency control or even phase synchronization of laser radiation. In the laboratories, systems based on semiconductor lasers and on diode-pumped Nd:YAG lasers were operated under frequency shift keying as well as with phase shift keyed modulation. Design studies performed in Europe and in the USA exemplify that coherent optical space links are the proper choice for a second generation, high-data-rate intersatellite communication system.

1. INTRODUCTION

At the time of this writing no laser-based communication link is operating in space. But remember - if you are old enough - that more than thirty years ago the invention of the laser was cheered as the ultimate solution for mankind's communication needs. Why this discrepancy?

Along the surface of the earth, glassfiber-based communications has become the stronghold of all cable-bound systems, it has even conquered the long-distance trans-oceanic routes. In space, the electromagnetic carrier employed to connect satellites is still microwave radiation, albeit a constant tendency to smaller wavelengths can be observed. Experimental satellites already use the frequency band of 32 GHz or even higher, corresponding to wavelengths in the centimeter region and below.

Designing laser communication systems for space has already started in the mid sixties. If present plans will indeed be realized, it will take another year until the first, low-capacity intersatellite link will be established. (This system /1/ is based on Nd:YAG lasers emitting at a wavelength of λ=1064nm and shall connect two geostationary satellites separated by some 80 000 km, at a data rate of 1.3 Mbit/s). Another experimental system is planned by the European Space Agency (ESA). This optical link will transmit a data stream of 50Mbit/s generated on board of a low-earth-orbiting observation spacecraft to a geostationary satellite. A semiconductor laser operating at λ=850nm will act as optical source /2/. The launching date is set for 1995.

From the technical viewpoint, the two systems just cited belong to the category of direct detection schemes. In their optical receivers, the only decision to be made is whether optical input power is incident onto a photodiode or not. No information on the exact frequency or phase of light is used. This detection technique is simple, but most primitive in the eyes of a communications engineer: It does not make efficient use of the wave nature of coherent electromagnetic radiation and therefore wastes bandwidth and transmitted power. Direct detection schemes are also susceptible to impairments due to background radiation and to technical noise.

If we look into the history of radio frequency communications, we have to go back to the time of Marconi at the turn of the century to find the corresponding primitive reception technique (see Fig. 1). This first generation of radio receivers was ousted in the thirties by heterodyne-based reception, a scheme still in use in today's radio and television sets as well as in microwave satellite links. The improvements are brought about by the inclusion of a so-called local oscillator in the receiver. Mixing its signal with the radiation to be received allows for optimal signal processing and results in a system efficiency closer to the limits set by information theory.

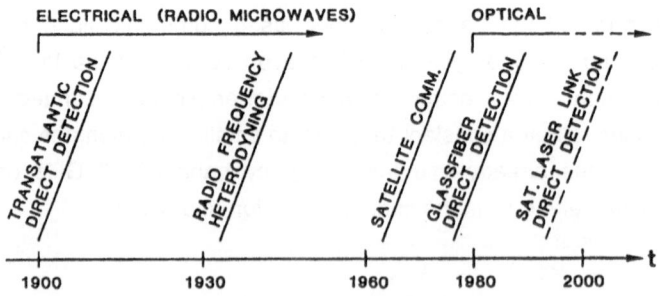

Fig. 1: Development of long-distance communications.

Can we expect a similar development in laser space communications? Is there room for a second generation system employing optical heterodyne or homodyne reception - generally termed coherent reception - and what will be the benefits? The answer is a clear "yes", though not even the first generation is up in the sky.

The main drivers for looking already for improvements is the meager input-power sensitivity of direct detection systems and their pronounced performance deterioration if strong background radiation enters the receiver's field of view. Link capacity of a free-space system can be increased by enlarging the transmit and/or the receive antenna diameter. This becomes evident from the well-known equation

$$P_R = P_T \, A_T \, A_R / (\lambda z)^2 \tag{1}$$

which expresses the received power P_R as a function of the transmitted power P_T, of the effective areas A_T and A_R of the transmit and receive antennas, of wavelength λ and of the distance z to be bridged. The required received power P_R increases with data rate. Hence, for given transmit power P_T, the distance and/or the data rate can be increased if the antennas are made larger. However, enlarging antenna size entails increased terminal mass and would weaken the competitiveness of optical space communication as compared to its microwave counterpart. The introduction of coherent optical receivers tackles the problem by providing better receiver sensitivity and hence asking for a smaller received power P_R.

Several expectations can be put forth when suggesting second-generation optical space links. One is the just-mentioned reduction in antenna diameter and hence mass. Figure 2 compares laser systems based on direct detection and on heterodyne detection and also includes microwave alternatives. Clearly, optical systems based on coherent heterodyne receivers fare best. The improvement over microwave systems is mainly a result of the reduced wavelength, as evidenced in equ. 1.

Another advantage lies in the possibility to employ a variety of modulation formats. While the direct detection scheme is limited to intensity modulation, the more efficient concepts of frequency or of phase modulation may be employed in connection with coherent receivers. This allows a choice of modulation format optimally harmonized with the laser type in the transmitter. Still another potential benefit is the excellent frequency selectivity of coherent receivers. They can be tuned to closely spaced channels, a property taken for granted in radio frequency engineering. A channel spacing of 1GHz or even less can be achieved with optical heterodyne receivers while for direct detection a spacing of less than some 5000GHz is not feasible. Therefore the available bandwidth

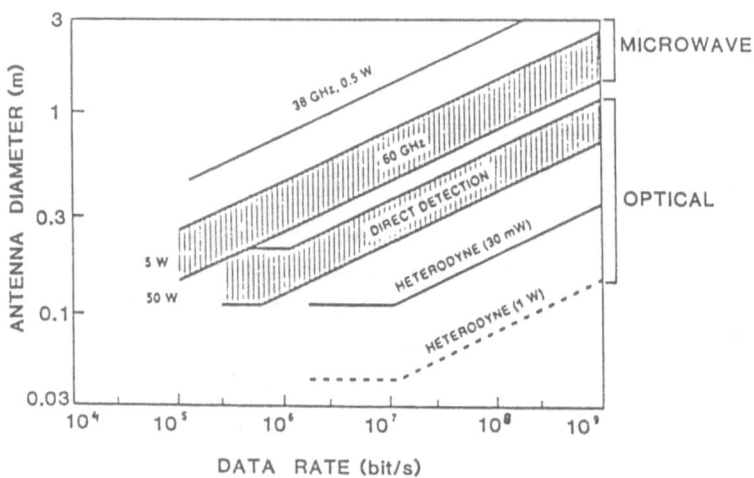

Fig. 2: Required antenna diameter for various ISL crosslinks (z=70 000km) as a function of data rate. Numerical parameters are transmitter power P_T and microwave carrier frequency. (After /13/).

can be used much more efficiently, an aspect which may play a role in future multichannel systems. (Note that in fiber optic communications the tendency to coherent systems is primarily caused by this unique property of coherent receivers).

The aim of this paper is to discuss features of coherent optical transmission systems useful for space-to-space links. I will concentrate on the problem of data reception and will only touch upon the question of pointing, acquiring, and tracking of space laser terminals. The paper should give the reader a profound insight into the potential these schemes offer and provide him with knowledge needed to judge future developments in this area. For this purpose the next chapter will define typical requirements of missions suited for laser communication. It is followed by Chapter 3 on basic concepts of coherent reception, including the difference between heterodyning and homodyning. In Chapter 4 I will discuss realizations of receivers by means of blockdiagrams and report on the state-of-the-art of available devices. Chapter 5 covers the properties of basic building blocks. The limits of receiver sensitivity are presented in Chapter 6 using the measure photons per bit. Deviations therefrom in actual implementations are dealt with in Chapter 7 on sensitivity penalties. A summary and an outlook to future development will conclude the paper.

2. MISSION SCENARIOS AND REQUIREMENTS

Several scenarios have been put up for the application of laser links. One consists of two (or more) satellites in geostationary orbit (GEO). They provide a connection of earth stations A,B (see Fig. 3) which, because of their geographical position, cannot be connected by a single satellite. The intersatellite link (ISL) avoids a double-hop link which would result in increased signal delay time. For down-links and up-links conventional microwave technology is preferable because of the detrimental absorption optical radiation may suffer in the lower atmosphere. Another mission scenario is shown in Fig. 4. Assume that the low-earth-orbiting observatory satellite LEO produces large amounts of data to be transmitted to earth station B. Uninterrupted data flow is possible by relaying the data to geostationary satellites (GEO) employing laser links before down linking via a microwave link. The same reasoning applies if instead of the LEO satellite we consider a space station, a polar platform, the space shuttle, etc. Because of the unequal orbits involved, the term interorbit link (IOL) has been coined for this type of mission. A third scenario concerns deep space missions like a probe to Mars or Saturn. With the enormous distances involved, microwave links to the earth would require untolerably large antennas on board of the probe, if medium-to-high-range data rates (5Mbit/s and above) are to be transmitted. Equation (1) demonstrates that increased distance z can be compensated by reducing wavelength λ accordingly, assuming that the other parameters stay unchanged. Switching from microwaves (50GHz) to laser light (300 000GHz) thus increases the maximum link distance by a factor of 6000.

For near-earth applications the most probable distances to be bridged are in the range between 40 000km and 70 000km /3/. These numbers result from the geostationary orbit distance of some 36 000km and for a symmetric arrangement of three GEO satellites, respectively.

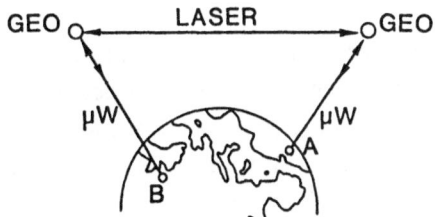

Fig. 3: Intersatellite laser link (ISL) between two geostationary satellites (GEO) to connect earth stations A and B. (μW...up and down link via microwaves).

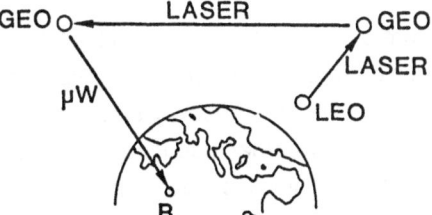

Fig. 4: Interorbit laser link (IOL) between a low-earth-orbiting (LEO) and a GEO satellite. An additional ISL provides uninterrupted data flow to earth station B.

Intersatellite links (ISL) are of interest for commercial voice, data, and televesion transmission. They should therefore sustain a two-way connection with equal data rates in both directions. The data rates are presently estimated to be in the range from 100Mbit/s to 1000Mbit/s. On the other hand, an IOL link will in general be asymmetric with respect to data rate. The so-called forward direction (LEO to GEO) has to carry the data generated at the LEO spacecraft, which may amount to 50Mbit/s or more /2/. The so-called return path (GEO to LEO) only has to provide for station-keeping signals. This may result in data rates of only a few Mbit/s.

An effect often negligible in the microwave regime is that of point ahead. In a two-way link where the terminals show a relative transverse velocity, the receive antenna and the transmit antenna of each terminal must not have parallel axes but have to squint slightly. The reason is the finite velocity of electromagnetic radiation. In order to receive the signal which was transmitted some time ago from the opposite terminal and to direct the own message to a position where the opposite terminal will be after the signal transit time, a so-called point ahead angle has to be implemented between incoming and outgoing direction. Its value is, of course, independent of wavelength. However, in the optical regime it may easily amount to several times the beam divergence and hence has to be taken care of, while in the microwave region it is typically much smaller than the beam divergence. Point ahead angles of some 40µrad to 70µrad will occur in ISL and IOL links, respectively.

A further effect not so pronounced at microwaves and not at all in direct detection optical systems is the frequency shift due to the Doppler effect. Consider an IOL link. The relative velocity v_R between the LEO and the GEO satellite may amount to some 7km/s. The difference between received (f_R) and transmitted frequency (f_T) is given by

$$|f_R - f_T| \approx f_T v_R / c \qquad (2)$$

where c denotes the velocity of light. For the large f_T encountered in optics this leads to a Doppler shift of some $|f_R - f_T| = 7$GHz. While a direct detection receiver will not notice this change, the high frequency selectivity of a coherent receiver requires automatic tuning to the received carrier frequency f_R. This can, hopefully, be accomplished by tuning the frequency of the local oscillator laser. Another characteristic is the rate with which the Doppler shift changes. A typical number for an IOL link is 10MHz/s. Both the Doppler effect itself and its rate can be advantageously used for navigation. This may be of interest for a deep space link - but it can be effectuated only by employing coherent reception.

Optical free space communication links are sensitive to visible background radiation. The sources of such radiation can be the sun, the moon, the sun-lit earth or other intense celestial bodies like the Venus. The GEO-bound receiver in the forward link of an IOL link, e.g., will have the earth in its field of view most of the time. For an ISL link this situation will not occur, but twice a year the sun will appear behind each transmitting terminal. The degradation of link performance in case of background radiation will be discussed in Chapter 7. Suffice to say here that a coherent system will suffer a minor penalty, while a direct detection system might turn inoperable during solar conjunction.

If laser space links are to excel their microwave counterparts, they must be realizable with low mass and low power consumption. Generally accepted numbers are 50 to 100kg and 100 to 200W for a two-way terminal. These values are not easy to reach nowadays, especially with coherent systems, because of their increased complexity as compared to direct detection systems.

Maturity of technology and space qualification are, of course, a prerequisite in any space application. For operational use - in contrast to experimental or pre-operational use - a lifetime of 10 years is a typical requirement. The associated reliability for the entire optical terminal is on the order of 0.8.

In this chapter we have seen that some of the mission requirements have equal consequences for direct detection and for coherent systems while others (like the sensitivity to Doppler shift or the sensitivity to background) differ in that respect.

3.THE CONCEPT OF COHERENT OPTICAL RECEPTION

3.1 Nomenclature

This chapter shall put some light on the terms coherent, heterodyning, synchronous, etc. This is appropriate, as the word coherent can be understood in many ways - and indeed it is used differently by radio engineers and by the optical communications community in connection with data receivers.

If one speaks of a coherent source of electromagnetic radiation one indicates that the phase of the emitted wave can be predicted (almost) for all times and for all locations. This implies that the fields involved resemble monochromatic travelling waves with sinusoidal dependence on time and space. That property is taken for granted for oscillator-generated fields in the radio and microwave regime. In optics it is, however, provided only by some laser types - and not at all by incandescent lamps or by sun light.

Laser light with a spectral bandwidth Δf much smaller than the center frequency is considered as being coherent. (An electrical engineer would compare laser light and thermal light with a sinusoidal wave and noise, respectively. Another comparison uses the waves on a water surface: If you throw just one large stone you get a coherent wave, if you throw a handful of pebbles, the wave is incoherent).

In radio communications, a coherent receiver is one where the phase of the received carrier is used to recover the data modulated onto that carrier. A heterodyne receiver is not necessarily a coherent receiver: It employs mixing with a local oscillator signal but it may demodulate the information without recourse to the knowledge of phase. Indeed, our radios and TV sets are not coherent receivers. However, some top quality microwave space links do employ coherent reception.

In optics, the term coherent communication system is generally used with a different meaning. It indicates that a local laser is involved in the reception process, regardless whether information on the carrier phase is employed or not. Hence any optical heterodyne receiver is called a coherent optical receiver. If further distinction is required, the term synchronous (or asynchronous) demodulation indicates the use (or not use) of the phase.

3.2 Heterodyning and homodyning

Two basic versions of coherent optical reception are shown in Fig. 5. To point out the difference to direct detection, I have included this most simple scheme as well (Fig. 5a). In direct detection, the photodiode acts as an optical envelope detector, its output current i is

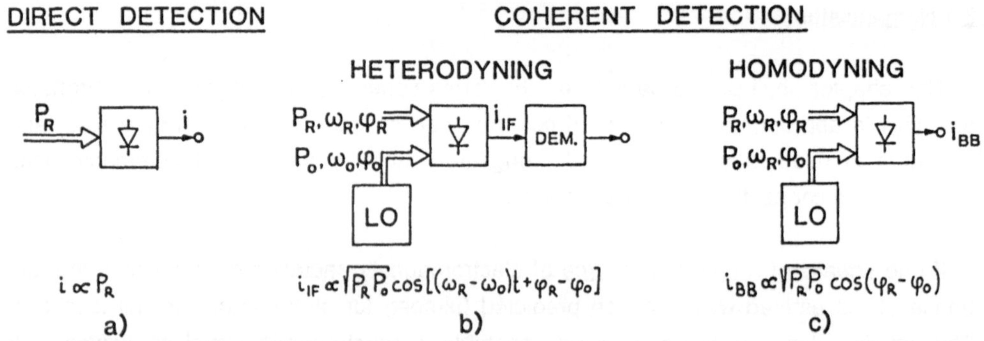

DIRECT DETECTION

COHERENT DETECTION

HETERODYNING

HOMODYNING

$i \propto P_R$

a)

$i_{IF} \propto \sqrt{P_R P_o} \cos[(\omega_R - \omega_o)t + \varphi_R - \varphi_o]$

b)

$i_{BB} \propto \sqrt{P_R P_o} \cos(\varphi_R - \varphi_o)$

c)

Fig. 5: Concepts for optical reception. (a) direct detection; coherent detection comprises heterodyning (b) and homodyning (c). P...optical power, ω,φ..frequency and phase of optical fields, IF...intermediate frequency, BB...baseband, LO...local oscillator, DEM...demodulation.

proportional to the optical signal power P_R. This process is negligibly dependent on the light's frequency, ω_R, and its phase, φ_R, and hence can detect only intensity modulation. With coherent detection, a strong local laser oscillator beam (power P_0, frequency ω_0, phase φ_0) is superimposed onto the weak received beam, the combined waves fall onto the photodetector. The photodiode current contains a component proportional to the product of the optical field strengths, $\sqrt{P_R}\sqrt{P_0}$, and to the cosine of the - in general time dependent - phase difference between the incident optical fields. Note, that increasing P_0 results in an amplification of the photodiode beat current. Further, not only amplitude variations but also frequency and phase variations of the input signal are reflected in the diode current. It follows that demodulation of amplitude-, frequency- and phase-shift-keyed signals (ASK, FSK, PSK) may be obtained.

In particular, Fig. 5b assumes ω_R to differ from ω_0. The input signal is then transposed into the intermediate band (IF) centered at $\omega_R\text{-}\omega_0$ before further processing (amplification, filtering, demodulation). It is this scheme which we call <u>heterodyning</u>.

If, on the other hand, the local oscillator frequency ω_0 is made to exactly coincide with ω_R, we speak of <u>homodyne</u> reception. The detector beat signal is then found in the baseband (BB), i.e. already demodulated. The requirement $\omega_0=\omega_R$ means that a phase synchronization to within a fraction of 2π has to be obtained - a task not easily achieved with every laser.

The coherent concepts of Fig. 5 ask for different electronic frequency bands to be provided by the stages following the photodiode. The heterodyne receiver requires a bandwidth twice the modulation baseband B with which the optical carrier (f_R) was modulated at the transmitter (see Fig. 6a,b). To allow for reasonably good realization of the electronic circuits, the intermediate frequency f_{IF} should exceed the bandwidth B by a factor of four. It follows from Fig. 6c that the homodyne receiver, because of $f_{IF}=0$, has relaxed spectral requirements: Including the photodiode, the unit has to cover only the bandwidth B. This reduces problems when realizing the unit, mainly with respect to circuit noise.

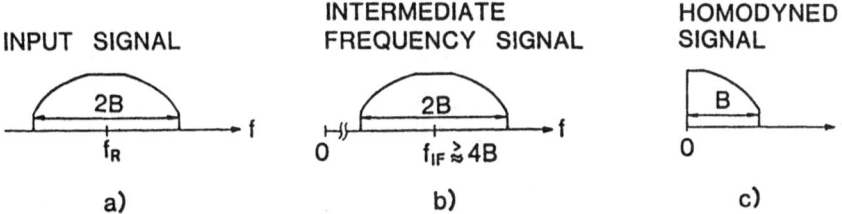

Fig. 6: Spectra in optical receivers. Input signal, baseband data width B << carrier frequency f_R (a). Intermediate frequency signal in the heterodyne receiver, $f_{IF}\gtrsim$ 4B (b). Homodyned signal (c).

3.3 Phase synchronization in homodyne receivers

While heterodyning is a well known technique - at least for communications engineers - the concept of homodyning can do with some further illustration. To this end I will discuss the phase relationship of the optical waves at the photodiode of a homodyne receiver in case of phase-shift-keyed (PSK) binary modulation. Figure 7 shows the corresponding phasor diagrams. The left sketch (Fig. 7a) represents full PSK where the phase difference between space and mark is $2\Delta=\pi$. To obtain maximum signal after photomixing, the phasor corresponding to zero and to the local oscillator have to be parallel (or antiparallel). As evident from the figure, the multiplying characteristic of the diode yields a photocurrent proportional to $\sqrt{P_R P_0}$ and $-\sqrt{P_R P_0}$ for a mark and a space, respectively. It is not a trivial task to maintain the indicated phase relationship because phase locking control loops basically yield a 90° relationship. One means to overcome this problem, the Costas loop, is discussed in Chapter 4. An alternative is to provide a residual carrier phasor symmetric to the sideband phasors by modulating with a phase deviation $\Delta<\pi/2$ (see Fig. 7b). The optical power of the residual carrier is $P_R \cos^2\Delta$, the optical power in the information-carrying sidebands is reduced by a factor of $\sin^2\Delta$ as compared to the ideal case. Another way to synchronize the local oscillator has been demonstrated recently /4/. It uses periodic time slots where, instead of space- or mark-encoded data, a full carrier phasor is transmitted. The scheme hence requires time multiplexing of data and pure carrier at the transmitter and corresponding demultiplexing at the receiver. The bit rate will be higher than the baud rate.

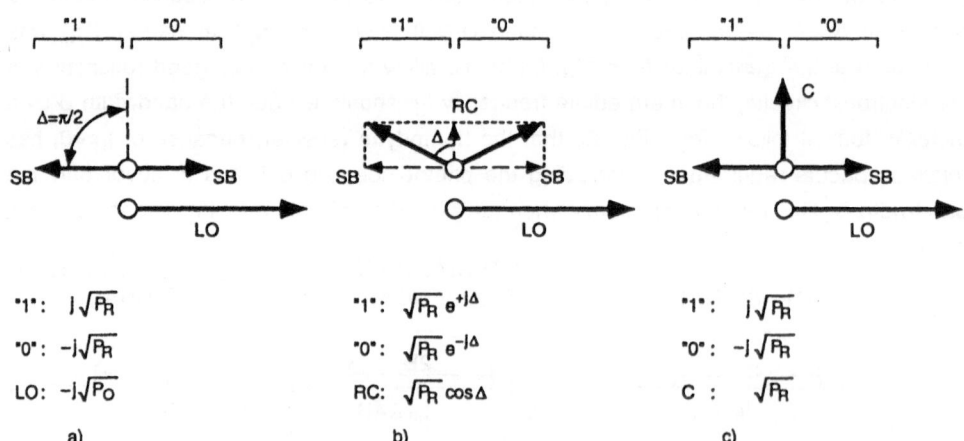

Fig. 7: Phasordiagrams for homodyne receivers with (a) full PSK, (b) residual carrier PSK, (c) PSK with synchronization bits. (Δ... phase deviation, SB...sideband, LO...local oscillator, RC...residual carrier, C...carrier, $j=\sqrt{-1}$).

3.4 Requirements for coherent reception

Compared to direct detection, coherent reception makes a number of additional demands. One is that the state of polarization (SOP) of the received field and the local field should coincide. In a space-to-space link the received SOP will most probably be circular to obtain independence of the terminals' relative orientation. This SOP is then changed into a linear SOP to coincide in type and direction with that of the usually linearly polarized local oscillator.

Another requirement for obtaining optimum beat signal is that amplitude distribution, phase front curvature, and wave direction of both fields at the photodiode should be identical. The usage of a fiber directional coupler for beam combining enforces this condition. However, losses will occur when coupling the waves into the fiber device. A second way of formulating this condition is the request for coaligned, identical modes. The local oscillator will, in general, be made to operate on the lowest-order transverse mode. The input field should match this mode closely and hence consist of a single mode as well. The telescope acting as a receive antenna in an optical data terminal must therefore be of high quality: It should introduce only negligibly small aberrations (e.g. wavefront distortions $<\lambda/20$), i.e. it has to work close to the diffraction limit. (In contrast, an antenna for a direct detection receiver only has to direct the incident power onto the photodiode and field distortions are not an issue. The term "photon bucket" has been coined for such a simple antenna). A consequence of the necessary alignment of mode directions is the narrow field-of-view of coherent receivers. If the input direction is off by more than the diffraction angle to be associated with the receiving optics, the electrical beat signal is heavily reduced. This property helps in discriminating against unwanted background radiation, but it reduces alignment tolerances and complicates the spatial acquisition process between two terminals.

A further, important requirement specific for coherent reception is one concerning the spectra of the fields to be heterodyned or homodyned. The ideal performance will only be obtained for monochromatic carriers. However, laser spectra do not exhibit negligibly small linewidths. Caused by spontaneous emission events, the optical field undergoes stochastic phase changes. The corresponding frequency noise is, in its simplest form, modelled as a white Gaussian process with zero mean. This leads to a Lorentz-shaped power spectrum. The linewidth of the beatnote resulting from the superposition of two lasers is the sum of both laser linewidths. In optical coherent receivers non-negligible linewidth is a more pronounced problem than in radio frequency receivers because it is the absolute value of the beat linewidth, Δf, which counts.

Depending on the receiver type, one can distinguish different mechanisms leading to increased bit-error-probability (BEP) due to phase noise /5/: In a homodyne receiver and in systems with differential PSK modulation (DPSK), phase noise blurs the phase relative to a reference, i.e. to the phase of the oscillator or that of the foregoing bit, respectively. If the phase to be determined is shifted by more than $\pi/2$, an error occurs which cannot be compensated by an increase of input power. The result is a BEP floor. In ASK and FSK heterodyne receivers with asynchronous detection, the phase noise manifests itself as a broadening of signal bandwidth. The intermediate frequency filter converts phase noise into amplitude noise which eventually leads to reduced signal-to-noise ratio. The resulting decrease in BEP, however, can be compensated by increasing the input signal level, albeit associated with a power penalty.

4. RECEIVER CONFIGURATIONS

Depending on the receiver concept (heterodyne or homodyne) and on the modulation format (PSK, FSK, ASK), various receiver configurations have been developed. In the following I will describe just two examples.

4.1 Homodyne receiver for PSK

As outlined in the preceding chapter, the main task in a homodyne receiver is phase synchronization of the local laser oscillator. Demodulation is performed automatically in parallel to the downconversion process. Let us assume a PSK-modulated input signal which contains no spectral line at the carrier frequency (see Fig. 7a). In this case phase locking requires a nonlinear circuit, such as the one known as Costas loop. Figure 8 shows the block diagram. An optical preprocessing unit combines the received field E_R and the local field E_O. This so-called 90° hybrid provides the input to two receiver arms. In the inphase arm, a photodiode generates a signal proportional to the cosine of the optical phase error φ. After regeneration, the demodulated data are available. The second arm produces the quadrature signal of the R and O inputs. A radio frequency mixer acts as a switching rectifier for the quadrature arm signal. Its output is low-pass filtered by the loop filter, yielding a control signal to be applied to the local oscillator. The loop tries to minimize the phase error φ. To achieve optimum BEP, the distribution of input power P_R among the two arms will be unequal, the quadrature (synchronization) arm requiring in general less power than the inphase (data) arm. The absolute power and the loop bandwidth required for successful synchronization depends on the a priori frequency stability of the lasers involved and on the accepted phase error standard variation σ_φ. A value $\sigma_\varphi < 3°$ will ensure negligible deviations from ideal performance. This receiver scheme combines the advantage of a small required electronic bandwidth in both arms,

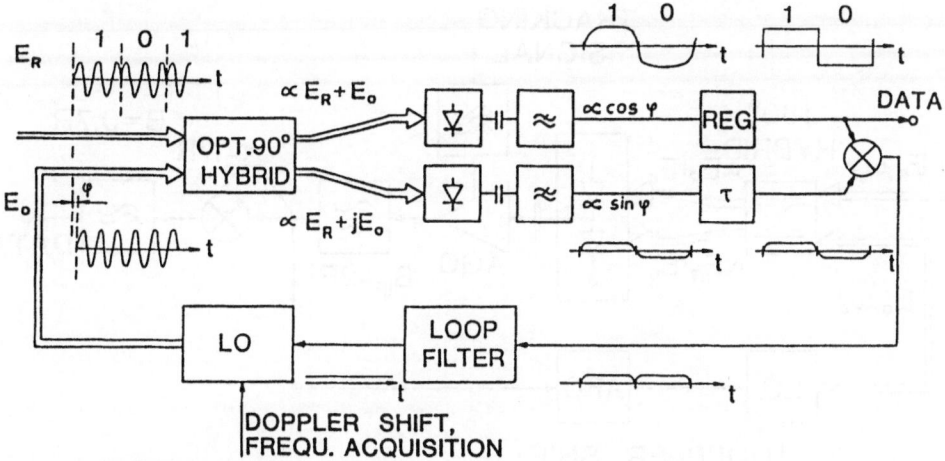

Fig. 8: Block diagram of a Costas-type PSK homodyne receiver. (REG...regenerator, LO...local oscillator).

being approximately equal to the data rate, and of ac-coupling of the control loop up to the mixer.

4.2 Heterodyne DPSK receiver

As one example for a heterodyne receiver I present the block diagram for DPSK modulated signals (Fig. 9). The received field E_R, ω_R is combined with the local field E_0, $\omega_0 \neq \omega_R$ in an optical device acting as so-called 180° hybrid, which produces both the sum and the difference of E_0 and E_R. In one implementation, this device may be a simple beam splitter, alternatively it may consist of a fiber directional coupler. The beat signals of the two photodiodes have different signs. In the arrangement shown, only the photocurrent difference feeds the following amplifier. Therefore the intermediate frequency signal centered at $|\omega_R-\omega_0|$ is processed optimally, but any terms resulting from direct detection are compensated. As a consequence, intensity noise due to varying local oscillator power, $P_0(t)$, will not deteriorate the system. After automatic gain control and IF-filtering, an autocorrelation demodulator performs asynchronous demodulation. This simple demodulation scheme asks for differentially encoded phase modulated input signals: A phase shift of π between successive symbols represents the transmission of a "mark", unchanged phase means the transmission of a "space". The automatic frequency control circuit (AFC) drives the local oscillator so to obtain an intermediate frequency of a multiple of half the data rate. The latter is required for proper delay time demodulation. Figure 9 also indicates that the AGC control signal may be used as input signal for the spatial tracking loop. For that purpose we assume a monopulse tracking scheme where

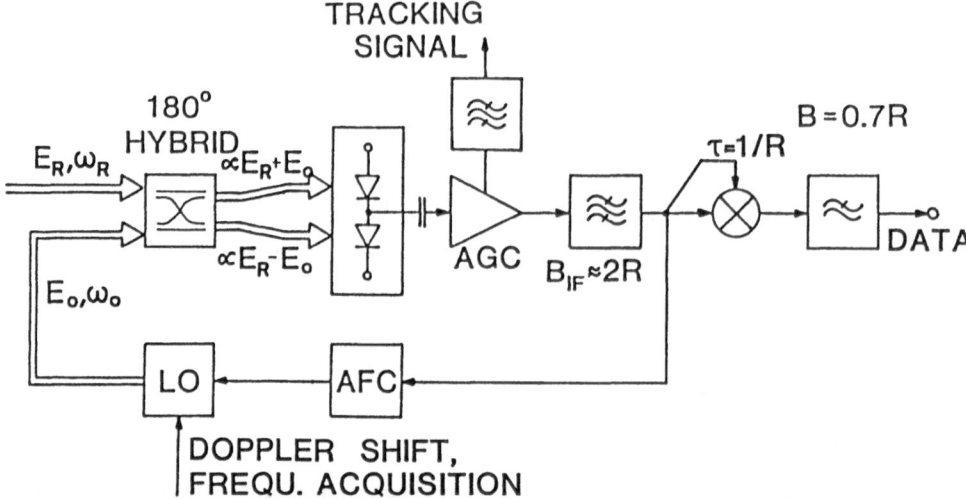

Fig. 9: Block diagram of a PSK heterodyne receiver with asynchronous demodulation, assuming DPSK modulated input signal. (AGC...automatic gain control, AFC...automatic frequency control, R...Data rate). A signal is provided for spatial tracking of the input direction (see text).

the receiver input beam is slightly scanned in a circular pattern around its optimal coupling position. The tracking loop tries to symmetrize the AGC voltage. The figure also shows that a second, coarse tuning input of the local oscillator may serve during frequency acquisition and assist in Doppler shift compensation. A DPSK heterodyne receiver requires neither an optical nor an electrical phase-locked loop and turns out to be a robust system.

5. DEVICES FOR COHERENT SYSTEMS

While many devices needed for coherent optical systems will not differ grossly from those for direct detection systems, some of them deserve special attention. Below I will discuss such aspects concerning lasers, modulation, and beam combining.

5.1 Lasers

Lasers used in the transmitter and as local oscillator must operate single-frequency. Requirements on the frequency noise power-spectral density primarily concern the beat note resulting from mixing the (unmodulated) carriers (see Chapter 7.2). The instantaneous linewidth of the beat note is given by the sum of the instantaneous linewidths of transmitter laser and local oscillator laser, Lorentzian lineshapes presupposed. The low part of the frequency noise spectrum shows up as a long-term frequency drift. The associated center frequency stability should be high, but tunability is required. The latter is especially important in homodyne receivers where phase control is implemented via fast frequency tuning. In systems with Doppler shift, the local oscillator must provide sufficient tuning range and rate. For the transmitter laser, an output power on the order of 1W is desirable for typical IOL and ISL links. The local oscillator laser will do with a few mW.

To make use of the more efficient modulation schemes of FSK and PSK, the transmitter laser must be modulated accordingly. While FSK is in general obtained by internal laser modulation, PSK may be obtained by external devices based on the electrooptic effect.

Presently two laser types are envisaged for coherent space communications, namely the diode laser and the diode-pumped Nd:YAG laser. The former offers internal FSK, high efficiency, and small volume, but it still lacks the cited transmitter output power when small linewidth is asked for as well. The Nd:YAG laser provides very small instantaneous linewidth (<1Hz) but suffers from low efficiency. It requires an external, quite power-consuming modulator.

5.2 Optical preprocessing

Optical preprocessors of the 180° hybrid type are readily available in the form of fiber-optic directional couplers. Advantageously they are based on polarization maintaining single-mode fibers. The 180° phase relationship is provided, as a gift of nature, by any conceivable low-loss four port device. And indeed, internal losses are sufficiently low (<0.1dB). However, stable low-loss coupling of freely propagating beams into the device is still an engineering problem. The realization of the 90° hybrid needed in Costas loop receivers (see Fig. 8) is not straightforward. A six-port device consisting of a beam splitter and two polarization separators has been developed, but presently only a bulk optics version has been tested /6/.

6. LIMITS OF RECEIVER SENSITIVITY

A key characteristic of a communications receiver is the minimum required input power, P_{min}, yielding a specified data quality. It may be stated in Watts or, equivalently, in decibels relative to 1mW (dBm). For digital systems, the data rate (R) and the bit-error-probability (BEP) should be given, too. (In general, P_{min} increases with increasing R and with decreasing BEP). With sensitivity comparisons in mind, however, it is more convenient to express sensitivity in average number of photons per bit (p). Then the data rate R is only of secondary importance.

The limit of sensitivity of optical receivers is given by the quantum nature of light. Deviations therefrom maybe due to nonideal photodiode quantum efficiency or due to other sources of noise such as circuit noise caused by the receiver electronics (shot noise due to diode dark current, thermal noise, transistor noise) and laser phase and intensity noise (see Chapter 3). If one succeeded in eliminating all these noise contributions, the so-called quantum noise would remain as the ultimate limit. This is in contrast to the microwave region where thermal noise processes dominate. This becomes clear from the well-known expression for the spectral noise power density S_N of an ideal amplifier (and also of an ideal detector or an antenna) given by /7/

$$S_N = hf/[exp(hf/kT)-1] + hf . \tag{3}$$

Here h and k are Planck's and Boltzmann's constants and T is the absolute temperature. Equation (3) is applicable throughout the electromagnetic spectrum. The first term represents thermal noise, the second term the quantum noise. For the optical region, hf>>kT and equ. (3) reduces to $S_N=hf$, while for microwaves hf<<kT and consequently $S_N=kT$.

The sensitivity limits for coherent optical receivers depend on the receiver concept, on the modulation format, and on the demodulation scheme. Table 1 gives equations for BEP as a function of photons per bit, p, and also states p for BEP=10^{-9} /8/. As shown, the best possible binary system is homodyne PSK. It asks for p=9 photons/bit to achieve BEP=10^{-9}. (For a data rate of R=100Mbit/s and a wavelength of 1µm this corresponds to P_{min}=0.18nW or -67.5dBm).

Table 1: Receiver bit error probability BEP expressed as a function of photons/bit, p. The lower entries give p for BEP=10^{-9}. (Asynchronous heterodyne PSK systems are usually realized via differential PSK; homodyne FSK is not feasible. For ASK, the number of photons per mark equals 2p). (After /8/).

	HOMODYNE	HETERODYNE	
		SYNC. DEMOD	ASYNC. DEMOD
PSK	$\frac{1}{2}$erfc$\sqrt{2p}$ 9	$\frac{1}{2}$erfc$\sqrt{p}$ 18	$\frac{1}{2}$exp($-p$) 20
FSK	—	$\frac{1}{2}$erfc$\sqrt{p/2}$ 36	$\frac{1}{2}$exp($-p/2$) 40
ASK	$\frac{1}{2}$erfc$\sqrt{p}$ 18	$\frac{1}{2}$erfc$\sqrt{p/2}$ 36	$\frac{1}{2}$exp($-p/2$) 40

Synchronous heterodyne systems are by a factor of 2 less sensitive than the corresponding homodyne systems. This is in contrast to the microwave region where both schemes offer equal sensitivity. The reason for this different behavior can be understood when again taking into account the unequal nature and location of the ultimate noise contributions. In the ideal optical coherent receiver, the noise power may be traced back to the shot-noise density in the photocurrent resulting from the local oscillator power. The total noise power is obtained by multiplying this density with the circuit bandwidth. The latter is equal to the baseband width B for homodyning and twice as much for heterodyning (compare Fig. 6). However, at microwave frequencies, thermal noise dominates, which is due to the antenna or preamplifier noise temperature. A corresponding noise density already exists at the input of the device which mixes received signal and local oscillator. Homodyne action may be considered as folding the noise density at frequency zero. This increases the density at the mixer output by a factor of two. The total noise power in the homodyne case (bandwidth B) is equal to the noise power in the heterodyne case, where density is not doubled but bandwidth is twice as large (2B). Therefore homodyning at microwave frequencies is attractive only because of small receiver bandwidth while at laser frequencies it additionally provides a 3dB sensitivity improvement.

It is known that the absolute value of the sensitivity limit in the optical regime is larger by two orders of magnitude than that for microwaves. This fact agrees well with a statement found in /8/ and with Table 1. It says that to detect a bit of information it requires about 10hf or 10kT of energy, whichever is greater. For f corresponding to 1µm and T=300K, the numbers are 2×10^{-18}Ws and 4×10^{-20}Ws.

Lastly one may ask about the sensitivity of an ideal direct detection receiver. The knowledge that light is Poisson-distributed suffices to derive BEP as a function of the number of photons per bit, p_{dir} /9/. One obtains

$$BEP = [\exp(-2p_{dir})] / 2 \qquad\qquad (4)$$

which yields $p_{dir}=10$ for BEP=10^{-9}. This value compares well with the results for coherent receivers. However, approaching this limit seems to be extremely difficult in practical realizations because of the various noise processes mentioned before. On the other hand, the limits of coherent receivers can be approached quite well by applying sufficiently large local oscillator power.

7. POWER PENALTIES

The ideal, quantum-limited sensitivites cited in the preceding chapter can not be obtained in real-world implementations. To achieve a prescribed sensitivity it may then be necessary to provide an optical input power $P_R > P_{min}$. Some effects, however, may lead to a BEP floor in the BEP/P_R dependence. Then even increasing P_R above all limits will not help in obtaining a prescribed (low) BEP. I have given examples for both cases in Chapter 3. In the following I will discuss effects and deficiencies leading to reduced sensitivity. Those which occur equally in direct detection receivers, like photodiode efficiency smaller than one, radio frequency circuit imperfections, intersymbol interference, and imperfect regeneration, will not be treated.

7.1 Local oscillator power-requirements

The ability of coherent receivers to approach the quantum-limited sensitivity is due to the amplifying property of adding the local oscillator power P_O (see Fig. 5b,c). Shot noise due to the local oscillator power increases as well. However, other noise power contributions, like circuit noise, may stay constant and can thus be neglected if P_O is made sufficiently large. The power signal-to-noise ratio can be written as /10/

$$SNR \propto P_R P_O / (2eSP_O + i_c^2) \qquad (5)$$

were e in the electron charge, S is the photodiode responsitivity, and i_c is the rms circuit noise current density. The numerator represents the amplified signal, the first term of the denominator gives the shot noise. Usually $P_O >> P_R$ holds and thus the shot noise due to P_R has been neglected in equ. (5). If we term the ultimate signal-to-noise ratio SNR_∞ - obtained for $P_O \to \infty$ - equ. (5) writes as

$$SNR = SNR_\infty / [1 + i_c^2 / (2eSP_O)] . \qquad (6)$$

This function is shown in Fig. 10. A power penalty of 0.5 dB or less is ensured if $2eSP_O / i_c^2 > 8.2$. (An example with $i_c = 5pA/\sqrt{Hz}$ and $S = 0.8A/W$ then asks for $P_O > 0.8mW$). In practical receivers the current noise density i_c increases roughly proportional to the operating frequency. Therefore heterodyne receivers require higher P_O than homodyne receivers if equal data rates are assumed. An upper limit of P_O may be either set by the availability of local laser power or by the maximum rating of photodiodes. In summary, even in coherent receivers low-noise receiver front ends are of prime importance if the quantum limit is to be approached.

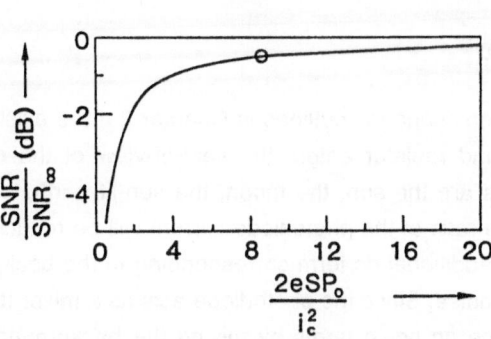

Fig. 10: Degradation of signal-to-noise ratio (SNR) due to insufficient local oscillator power P_o. (e...electron charge, S...photodiode sensitivity, i_c...circuit noise current density).

7.2 Laser phase noise

The phenomenon of laser phase noise has already been mentioned in Chapter 5. A complete characterization of phase noise is done by stating the corresponding frequency noise spectral density. If its frequency independent (or white) portion dominates, the resulting laser linewidth is an equivalent description. In this case the relevant quantity for the deterioration of sensitivity of a coherent receiver is the linewidth of the beat note, given by the sum of the transmitter and the local oscillator linewidth. A number of theoretical analyses state the maximum tolerable beat note linewidth, Δf_m, if a certain penalty in receiver sensitivity is tolerated. Table 2 gives some examples. It turns out that the more sensitive a system is, the stronger it also feels the influence of non-zero linewidth. What counts is the ratio of linewidth and data rate. This is not surprising: It says that it is the phase coherence within one bit duration which is relevant.

Table 2: Maximum tolerable beat linewidth, Δf_m, normalized to data rate R for various coherent optical communications schemes. The cited values either lead to a BEP floor of 10^{-10} or they imply a 1dB power penalty |5|. For the FSK scheme the value depends on the frequency deviation of the received signal.

	PSK			FSK	ASK
	HOMODYNE	HETERODYNE		HETERODYNE	HETERODYNE
		SYNCHRON.	ASYNCHRON.	ASYNCHRON.	ASYNCHRON.
$\Delta f_m/R$	6×10^{-4}	4.5×10^{-3}	3.3×10^{-3}	0.02 to 0.09	0.18

7.3 Background radiation

For each of the mission scenarios outlined in Chapter 2 there exists a probability that a non-negligible background radiator enters the field-of-view of the optical receiver. The most prominent sources are the sun, the moon, the sun-lit earth, and Venus. In such a case, the signal-to-noise ratio of the photodiode current will be reduced. The photocurrent will not only contain an additional dc term corresponding to the background power, which leads to additional shot noise. Since the photodiode acts as a mixer for the incident optical fields, it will also produce ac noise terms by mixing the background field with itself and with the data input signal. In the case of heterodyning or homodyning, mixing with the local laser must be taken into account as well. As a consequence of these noise contributions BEP rises, and a power penalty is inferred if BEP has to be kept at its initial value. Depending on the system parameters (receiver antenna of the optical terminal, type of receiver, background source, etc.) degradation ranging from negligible to disastrous may occur.

In direct detection receivers it is usually the additional photocurrent dc term which dominates. Because of the heterodyne/homodyne process, a coherent receiver provides both an excellent spatial and spectral filter function, thus being less vulnerable to background light. Some influence, however, always remains. It is the influence of the cross term between local oscillator power and background signal which is decisive. Consequently, increasing the local oscillator power will bring no reduction of background effects. Figure 11 gives examples for the degradation DEG (in dB) of the signal-to-noise ratio of a coherent receiver as caused by sun radiation /11/. The abscissa value β is the a priori deviation of receiver performance from the shot noise limit, η is the photodiode quantum efficiency, λ is the wavelength. The influence of background power is felt more strongly for a good coherent receiver approaching optimum performance ($\beta \rightarrow$ 0dB). A

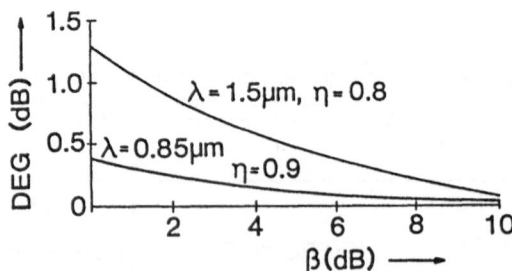

Fig. 11: Degradation of coherent reception due to sun radiation. The parameter β measures the receiver's sensitivity relative to the quantum limit in case of zero background (η...photodiode quantum efficiency).

typical value for the degradation even in case of sun background is as low as 0.5dB. This must be seen in comparison with the corresponding value for a direct detection receiver where the degradation may amount to some 10 to 20dB.

7.4 Heterodyne efficiency

Within Chapter 3 I have already mentioned that a coherent receiver achieves the highest possible beat note signal only in case of matched received and local fields. Matching is required with respect to polarization, transverse mode, field diameter, and field direction. Let us assume the likely case that field combining takes place in a singlemode directional coupler. To make best use of the valuable power intercepted by the terminal's telescope antenna, one aims for the highest possible degree of coupling into the coupler's input port. The telescope output beam will be focussed onto this fiber end. The resulting field distribution is that of an Airy pattern. Even under optimum adjustment of its diameter, the overlap with the fiber's eigenmode is smaller than unity, as the eigenmode's field approximates a Gaussian distribution. This unavoidable mismatch results in a coupling loss of 1dB. It represents the heterodyne loss if we neglect the usually very small losses due to polarization mismatch.

Another loss contribution one must take into account is due to wavefront distortion within the telescope. Any deviation from ideally shaped surfaces of the optics involved distorts the Airy pattern at the fiber end and thus further reduces the coupling efficiency. Analysis shows that an rms wavefront distortion of $\lambda/20$ will result in additional coupling losses of 0.4dB. (For $\lambda/13$ and $\lambda/40$ one obtains 1dB and 0.2dB).

7.5 Sensitivities achieved in breadboard receivers

Mostly driven by the development in coherent fiber communications, a large number of receiver implementations have been reported within the last years. Most of them were based on semiconductor lasers, some on diode-pumped Nd:YAG lasers. Figure 12 shows the achieved sensitivites for BEP = 10^{-9} in photons per bit as a function of data rate. It covers heterodyne and homodyne receivers and, for comparison, also includes cases of direct detection. The theoretical limits of 9 photons/bit (homodyne), 10 photons/bit (direct), and 18 photons/bit (heterodyne) are given as horizontal lines.

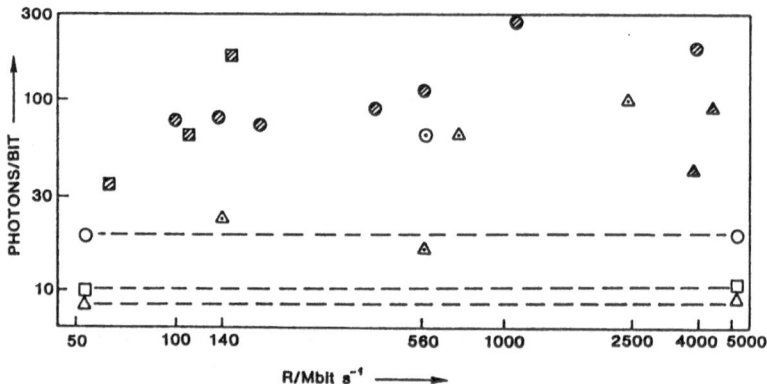

Fig. 12: Reported sensitivities of coherent receivers at BEP=10⁻⁹ as a function of data rate
R. For comparison, three values obtained with direct detection are included. The
symbols O, △, and ☐ relate to heterodyne, homodyne, and direct detection, the dot
and the hatching indicate Nd:YAG and diode laser systems, respectively. The
dashed horizontal lines indicate the quantum limits.

8. SUMMARY AND OUTLOOK

The concept of coherent reception has proved itself as extremely powerful in
transmission links based on radio and microwave frequencies. There is no doubt that the
same principle will be equally efficient and advantageous in the optical regime. Compared
to the more primitive direct detection, coherent reception in laser links will benefit from

- input power sensitivity close to the quantum limit,
- the possibility to employ phase or frequency modulation formats,
- reduced sensitivity to background radiation,
- reduced sensitivity to interference in a duplex link,
- a more economical use of channel bandwidth,
- increased frequency selectivity and hence the possibility of multichannel transmission,
- higher dynamic range due to the photocurrent being only proportional to the square
  root of the optical input power.

These advantages do not come for free. The necessity to provide a local laser radiation
of proper frequency and/or phase relationship and the need of painstaking superposition
with the input field entails higher receiver complexity and tighter device specifications. In
particular I recall the requirements for

- high spectral purity of transmitter and local oscillator laser,

- a local oscillator laser with
  * sufficiently wide and fast tunability to cover transmitter frequency instabilities and Doppler shift,
  * sufficient output power to obtain (near) quantum-limited receiver operation,
  * acceptable intensity noise,
- frequency and/or phase acquisition before data can be received.
- a low-loss device for combining input and local field with proper phase relationship,
- polarization matching of the fields to be combined.

Further, the telescope serving as receiving antenna must be of near-diffraction-limited quality. The desired intermediate frequency signal will be produced only if the single-moded local oscillator field is superimposed by an input field of the same spatial mode, but not by orthogonal modes. Therefore the field-of-view of the receiver terminal is very small, namely given by the antenna's diffraction limited divergence.

I maintain that there are no basic obstacles to arrive, even today, at a well-engineered coherent optical receiver. However, to demonstrate reliability and space qualification will ask for additional efforts. This specifically applies to the lasers, the external modulator, and to alignment and coupling elements. To achieve a more pronounced weight advantage over microwave terminals, efforts are required to obtain lower terminal mass than that presently envisaged (i.e. some 70kg). (Ten years ago it has been estimated that a laser crosslink subsystem is lighter than a 60 GHz microwave system only for data rates exceeding 1Gbit/s /12/). In short, it will still take some development time and cost before we can expect a coherent optical data link in space.

System designs and hardware pre-development have reached high maturity. In closing this paper I want to shortly characterize just two projects. One was devised at the MIT Lincoln Laboratory /12,13/. It uses GaAlAs diode lasers with an output power of 30mW and a four-fold redundancy in the transmitter. The lasers operate at λ=0.86µm, with a frequency settability of ±100MHz. The modulation format is quaternary frequency shifting with tone spacings of 220MHz, the data rate is 110Mbit/s. A heterodyne receiver built around a silicon pin photodiode achieves a sensitivity of 80 photons/bit at BEP=10^{-5}. The second system was designed within a contract of the European Space Agency (ESA) /3/. For the forward direction of a LEO-GEO link a homodyne system based on diode-pumped Nd:YAG lasers emitting at λ=1.06µm was chosen. The output beam with one Watt power is phaseshifted in an external LiNbO$_3$ modulator at a data rate of 650Mbit/s. A Costas loop homodyne receiver is designed to operate some 7dB above the quantum noise limit. This corresponds to an input power of 5nW at BEP=10^{-7}. A parallel study with similar results was performed by an Italian team /14/. Efforts are under way to realize, within the

German national space program, an engineering model of the entire laser terminal under the project name SOLACOS (solid state laser communications in space).

References

/1/ R.B. Deadrick, W.F. Deckelman, "Laser crosslink subsystem - an overview", to appear in Proc. SPIE 1635, (1992).

/2/ G. Oppenhäuser et al., "The European SILEX project and other advanced concepts for optical space communications, Proc. SPIE 1522, 2 (1991).

/3/ H. Sontag et al., "Design of a diode-pumped Nd-Host laser communication system", Final Report, ESTEC Contract No.8533/89, Dec. 1990.

/4/ B. Wandernoth, "1064nm, 565 Mbit/s PSK transmission experiment with homodyne receiver using synchronisation bits", El. Lett 27, 588 (1991).

/5/ L.G. Kazovsky, "Coherent optical receivers: performance analysis and laser linewidth requirements", Opt. Eng. 25, 575 (1986).

/6/ W.R. Leeb, "Optical 90° hybrid for Costas-type receivers", El. Lett. 26, 1431 (1990).

/7/ B.M. Oliver, "Thermal and quantum noise", Proc. IEEE 53, 436 (1965).

/8/ P. S. Henry et al., "Introduction to ligthwave systems", in "Optical fiber telecommunications II", eds. S.E. Miller and I.P. Kaminow, Academic Press, 1988.

/9/ G. Grau, "Schwankungserscheinungen als prinzipielle und praktische Leistungsgrenzen optischer Nachrichtensysteme", A.E.Ü, 37, 137 (1983).

/10/ W.R. Leeb, "Heterodyne and homodyne detection in optical space communications", Proc. SPIE 1131, 216 (1989).

/11/ W.R. Leeb, "Degradation of signal to noise ratio in optical free space data links due to background illumination", Appl. Opt. 28, 3443 (1989).

/12/ V.W.S. Chan, "Space coherent optical communication systems - an introduction". J. Lightwave Technol. LT-5, 633 (1987).

/13/ V.W.S. Chan, "Intersatellite optical heterodyne communication systems", Proc. SPIE 1131, 204 (1989).

/14/ A.E. Marini, A. Della Torre, "Definition of interorbit link optical terminals with diode-pumped Nd:host laser technology", Proc. SPIE 1522, 222 (1991).

THE NASA PROGRAM FOR OPTICAL
DEEP-SPACE COMMUNICATIONS AT JPL

James R. Lesh
Jet Propulsion Laboratory
California Institute of Technology
4800 Oak Grove Drive
Pasadena, California 91109
(818) 354-2766

ABSTRACT

As deep-space missions to the planets and other
heavenly bodies acquire larger amounts of more
sophisticated data, the demand for greater data return
link communications capacity continues to increase.
Past and current missions have stretched the
capabilities of radio frequency systems close to their
fundamental limits. For these reasons, NASA has been
supporting the development of technologies and systems
designs that will allow future missions to use optical
beams for communications. Such systems will permit
dramatic increases in return link capacity with
substantially smaller and less massive communications
subsystems on board the spacecraft. The history and
past accomplishments of the JPL deep-space optical
communications program will be described, along with
the plans for the future development, demonstration,
and eventual deployment of this important technology.

I. INTRODUCTION

As mankind extends his reach farther into deep space his
thirst for knowledge continues to grow. Each new obstacle
overcome reveals yet another frontier to be explored; another
challenge to be met. Initial deep space missions explored the
closer regions of space (the Moon, Venus, Mars, etc). Some of the
earliest missions returned only minute amounts of data by
conventional standards with data rates as low as 8.33 bits/sec.
As time went on and technology matured, the ability to return
data at higher rates and from further distances was realized.
Current day missions can return as much as 115 Kbps from as far
as Jupiter and can send at more moderate rates from the far
extremes of the Solar system.

This expansion in communication capability has not come
without a substantial investment in technology and resources. The
Voyager mission, which has traveled beyond Neptune, carries a 3.7

meter radio-frequency (rf) antenna and uses a 20 Watt X-band traveling wave tube (TWT) for data return. When it encountered Jupiter, the data rate to one of the Deep Space Network's (DSN's) largest 70 meter diameter antennas was only 115 Kbps. Upon reaching Uranus, it was necessary to array-combine the outputs of several (DSN) antennas just to achieve a few 10's of kilobits per second, and upon reaching Neptune, the DSN antennas were combined with the 27-element Very Large Array in Socorro, New Mexico to maintain such a moderate data return.

With the Bold New Initiatives being considered by NASA the problem of insufficient data link capacity will become even more acute. New instruments like synthetic aperture radars (SAR) and high-resolution imaging spectrometers (HIRIS) will require as much as perhaps 100 Mbps each, and a number of future missions are considering such instruments for orbital missions to outer-planetary satellites. A piloted mission to Mars alone will require capabilities such as two-way video links and substantial telemetering to monitor and ensure the health of the crew.

To provide increased link capacity for these future missions, NASA is sponsoring research and development of laser communications technology at the Jet Propulsion Laboratory. The program commenced around 1980 with a small study effort to examine the theoretical limits of photon communication efficiency, and has expanded today to an entire group of full-time researchers working on technology development, systems design, and future deployment planning. When laser communications will actually be deployed operationally is not certain at this time. However, a series of experiments that will demonstrate the maturity of the technology are being considered. This paper describes the deep-space optical communication program currently underway, discusses the several technology development accomplishments that have been achieved, and identifies the currently foreseen challenges that must be overcome before operational use can be realized.

In section II we will discuss a number of the benefits of using optical communications as compared to conventional rf technology. Next, the anticipated systems designs and overall architectures for future optical communications systems will be described. In section IV we will discuss the technology developments which apply to the spacecraft end of the link. This will be followed by a similar section covering of the technology and systems design activities for the Earth reception end of the link. In section VI we will describe the long range development plan for this technology. Then in section VII we will summarize with conclusions.

II. BENEFITS OF OPTICAL COMMUNICATIONS

As mentioned in the introduction, optical communications offers a substantial increase in link capacity. This increase comes as a result of the much smaller wavelength associated with the optical carrier. Smaller wavelengths result in narrower transmitted beam divergences and hence more concentration of the power on the intended target receiver. For example, if one compares X-band (3 cm) rf with visible light (0.5 um), the ratio of power concentration is 95 dB. Not all of this gain results in link advantage however. The basic process of photodetection is less efficient than for rf due to quantum noise effects. When taking into account these effects, the resulting overall link benefit (which depends on the particular operating conditions of the link) is typically about 71 dB.

One might be tempted to assume that a 71 dB increase in link capacity would therefore be available. This is not the case since it would require transmitting and receiving apertures that are as large as the corresponding rf systems. To make a 4 meter diameter diffraction limited optical spacecraft transmitting telescope, for example, is clearly not practical from a cost standpoint, and the propulsion requirements for a spacecraft carrying such an aperture would be enormous. However, having 71 dB of advantage means that the design requirements on the various optical system components can be greatly reduced. This can result in much more realistic parameters and still provide a substantial increase in communication capability.

To illustrate this principle, let us consider the Voyager spacecraft with its 3.7 meter X-band antenna. From the mean Saturn distance of 10 AU (1 AU= 1 astronomical unit or the distance between the Sun and the Earth), the size of the propagating beam, by the time it reaches the Earth, is 1000 Earth diameters wide (see Figure 1).

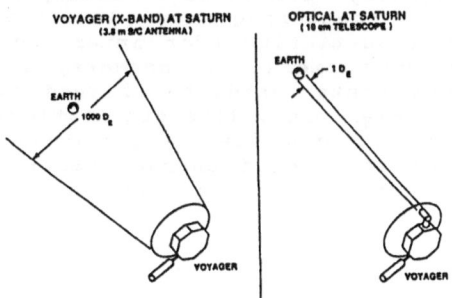

Fig. 1. Deep-Space Communications Beam Spread

Contrast this with a 10 cm telescope using visible (0.5 um) light from the same distance. The corresponding footprint at the Earth is only about 1 Earth diameter wide. Thus, a power concentration of 60 dB can be achieved using a much smaller transmitter aperture. That 60 dB gain can then be used to reduce the requirements on other parts of the system (transmit laser power, receive aperture diameter, relaxed pointing requirements, and, of course, the difference in inherent detectability), as well as increase the overall link capacity. Thus we see that the spacecraft hardware for comparable or better link capacity can be substantially smaller.

In addition to direct benefits of size, mass, shape and power, there are also secondary benefits that can be accrued by having an optical communication package on board a planetary spacecraft. Two of these are precise navigational tracking and serendipitous scientific opportunities. Each of these will be covered briefly.

Not only must we communicate with a deep-space vehicle to extract the desired science data gathered by it, we must also spatially track it for navigation and orbit determination purposes. The most challenging variables to obtain are the two angular position coordinates since the error in these quantities is multiplied by the range to the spacecraft. Currently the most accurate method for obtaining these quantities is by rf very long baseline interferometry. With optical signals, the signal from the spacecraft can be viewed (imaged) directly against the stellar background. Thus, both angle coordinates can be measured directly from a single optical telescope.

With conventional deep-space missions there is a scientific field that has developed around the use of radio-frequency signals for communication. This field is known as radio science. Here the radio communication signals are used to extract scientific information about the encountered planetary systems or the intervening space link medium. Optical signals enable an entirely analogous field of "light science". This field includes such things as planetary atmospheric absorption at optical wavelengths, fine-scale scattering from planetary ring systems, and integrated forward scattering over interplanetary distances in the Solar system dust field. Furthermore, many previous rf experiments which were contaminated by charged particle density fluctuations from the Solar wind, like gravitational body bending of electromagnetic waves, can now be performed without those disturbances since the effects of charged particle fluctuations fall off as the square of the carrier frequency.

III. SYSTEMS CONCEPTS AND ARCHITECTURES

A. Overall System

Initially, reception of signals from deep space could be performed by ground-based receivers. Early systems will use intensity-modulated signals with direct detection reception, so one need not worry about preserving spatial phase coherence while propagating through the atmosphere. Cloud cover will, of course, be a major impediment. To circumvent this, several spatially diverse reception stations operating in independent weather cells can be used. Studies have shown that three independent sites can

produce an overall visibility to the ground of 95% [Ref 1].
Realization of such a network could be as soon as the mid 1990's.

It is, however, envisioned that spaceborne reception will
eventually be used. There are two reasons for this. First, it
gets the receiver above the cloud cover, as well as the phase
front disturbances caused by atmospheric turbulence. Second, by
being outside the atmosphere, background light associated with
daytime sky scattering will be eliminated. Spaceborne reception
could become a reality in low Earth orbit, perhaps aboard or near
the Space Station, by the turn of the century. However, this
still leaves Earth blockage to a deep-space probe approximately
half of the 90 minute station orbit period. Thus, one would
really like to have such a station located in a much higher
orbit, perhaps in geosynchronous orbit, or at one of the stable
libration points. Such a system would likely not be deployed
before the year 2015. Once received by the orbital terminal, the
data would subsequently be relayed to the ground via conventional
rf techniques.

B. Spacecraft

The basic architecture for a planetary spacecraft optical
transceiver subsystem is shown in Figure 2.

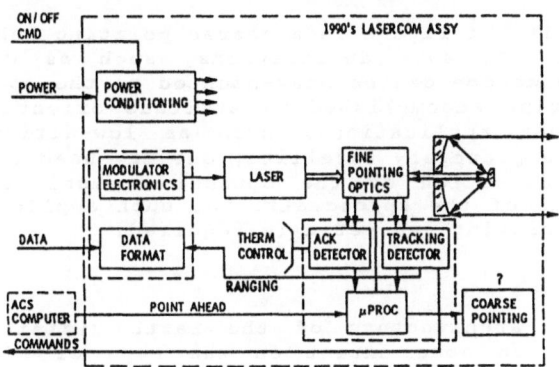

Fig. 2. Spacecraft Subsystem

In contains a single telescope, usually in the 10-50 cm diameter
range. This telescope is used for both receiving the beacon (eg
Solar-illuminated Earth) or other uplink signal transmissions, as
well as for transmitting the downlink signal. Telescope pointing
need only be accomplished to the 1-2 milliradian level
(comparable to the attitude control deadband cycling of a typical
planetary spacecraft) since the telescope's field-of-view is
large enough (perhaps 5 mrad) to include the intended target.
Fine resolution for target acquisition and tracking is
accomplished by means of a set of fine pointing optics operating
in conjunction with an acquisition detector array, with
subsequent hand-over to a tracking array. The spatial range of

these components is large enough to cover the field-of-view of the telescope. Additionally, selected uplink signals, in the form of command or ranging signals, can be detected by an appropriate reception detector.

Data collected by the spacecraft is applied to a data formatter and combined as appropriate with detected ranging signals for transmission to the Earth receiver. The formatted signal is applied to a modulator assembly for subsequent modulation of a laser. The output of the laser is then directed out through the telescope via a second set of fine pointing mirrors directed in such a way to point the beam back to the target being tracked. Because of the narrow beamwidths of optical systems, and due to the finite speed of light, optical communications signals must be pointed ahead of the apparent location of the intended target receiver so that the transmitted beam will intercept it when the signal reaches the receiver. The magnitude of this point-ahead angle depends on the relative cross velocity of the two communications terminals and can be as large as 500 microradians in some planetary applications. The transmit path steering mirrors are used to introduce this offset angle. Additionally, some portion of the transmit beam is usually picked off in the spacecraft and imaged onto the tracking detector for receive /transmit calibration purposes.

Also shown in the figure is a coarse pointing subsystem with a question mark. In some applications, such as outer planet missions, the telescope can be body-mounted to the spacecraft and its coarse pointing accomplished by attitude orientation of the vehicle. In other applications, such as low-altitude orbital missions around a planetary satellite, one may need to separately articulate the direction of the optical communication package from the attitude of the spacecraft. For such applications, some form of coarse pointing subsystem is required.

C. Receiver End

The overall architecture of the Earth vicinity reception terminal depends to some extent on the type of signaling and detection systems employed. For the near term, intensity modulation with direct detection is favored due to the resulting robustness relative to atmospheric turbulence, and the increased ease of receiver implementation for either ground-based or space-based operation. Beyond that, and particularly for space-based reception, coherent detection systems may be utilized. In this section we will concentrate on direct detection.

Figure 3 depicts the basic structure for the receiving terminal. The most significant ingredient is a large aperture telescope used for reception of the returned optical signals. Aperture sizes in the 1-20 meter diameter range have been studied, although the baseline size is 10 meters. The telescope, however, is not a high quality imaging system. Instead, it is a non-diffraction-limited photon bucket collector. Such a system does not preserve optical phase across the aperture, but instead simply collects the incident photons and concentrates them onto a small enough blur circle at the focal plane so that an appropriately wide field-of-view detector can measure their arrival.

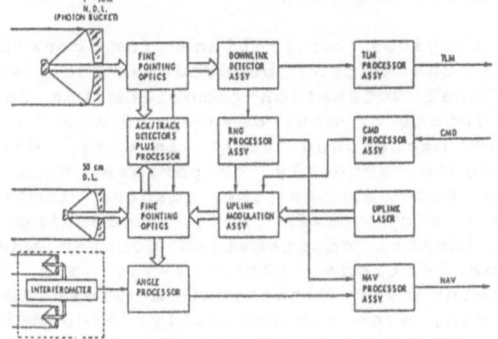

Fig. 3. Optical Reception Station

By not requiring the preservation of phase, the large aperture surface can be realized much more economically. Pointing of this telescope is also only required to the few milliradian level, as smaller fine pointing optics can be used, like in the spacecraft, to search over the telescope's overall field-of-regard. The primary surface is expected to be segmented to reduce fabrication and maintenance costs, and to permit eventual transfer of the technology to spaceborne reception systems.

A second smaller telescope, perhaps around 0.5 meters in diameter, is also envisioned as an uplink transmitting telescope. It is expected to be near diffraction-limited so that the emitted field has a reasonably good propagating wavefront. (Of course, if the system is on the ground, atmospheric turbulence will cause appreciable beam broadening). This telescope is driven by the uplink laser, coupled through an appropriate ranging or command modulation system.

The smaller telescope can also be used in a receive imaging mode to image the signal from the spacecraft against the stellar background. In such a mode, information about the spatial (angular) location of the spacecraft relative to the stellar inertial reference frame can be obtained. This information, coupled with turn-around range delay information, is important in determining the trajectory of the spacecraft for navigational purposes. In the future, and particularly for space-based reception, such an angle measuring capability can be further improved by the use of wider baseline interferometric techniques.

IV. SPACECRAFT AND LINK TECHNOLOGY DEVELOPMENTS

There are a number of key technological ingredients that

have been developed over the years to support deep-space optical
communications. Several of these key developments will now be
described.

A. Multi-bit/Photon Communication

        One of the original motivations for considering optical
communications was the narrow beamwidths that are achievable.
However, an additional motivation came from the realization that
under conditions typical of many deep-space missions (ie moderate
data rate and low background light levels), direct detection
optical systems could actually outperform optical heterodyne
detection, and in fact exceed the quantum limit in terms of
signal power efficiency [Refs 2-4]. Accordingly, a design,
analysis and experimental confirmation program was initiated to
confirm these projections [Refs 5-7]. The quantum limit
associated with heterodyne detection, a limit beyond which no
heterodyne system can, even theoretically, exceed is 1 nat/photon
[Ref 3]. Since 1 nat is approximately 1.44 bits of information,
the demonstration goal was set at 2.5 bits of information per
detected photon (approximately twice the heterodyne limit).
        The basic concept is shown in Figure 4.

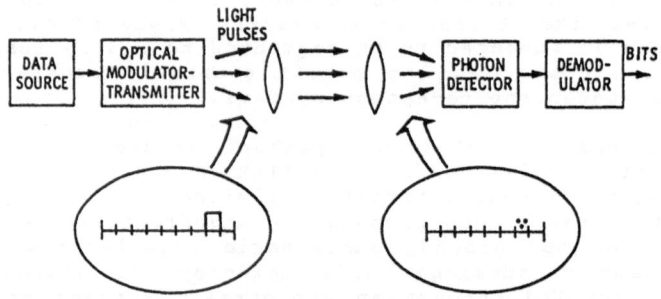

Fig. 4. Multi-bit/photon Concept

At the transmitter end, a time window of length T, seconds is
divided into M equal-length slots. Information is transmitted by
placing a pulse of light in only one of the available slots in
the T second interval. Since the location of the pulse can be
represented by log (M) bits, then one pulse can specify that
number of bits of information. In the figure, M=8, so that one
pulse can convey 3 bits of information. Now assume that the
pulse, after propagating over the link, is received with an
intensity on 3 photons (see figure). If that pulse can be
reliably detected, the reception of three photons can convey 3
bits over the link. In the demonstration set-up, M was actually
256 (representable by 8 bits) and the corresponding signal
intensity was about 3 photons (8 bits/3 photons = 2.67
bits/detected photon). Special care must be taken to optimize the

probability of detection at such weak signal levels and certain erasure fill-in codes must be used to compensate for the probability of missed detection due to quantum fluctuations.

The system used to demonstrate such performance is shown in Figure 5.

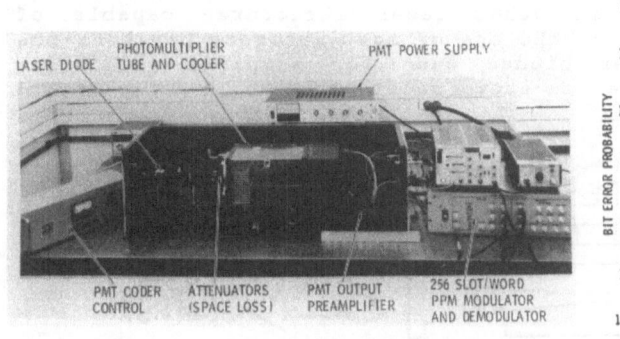 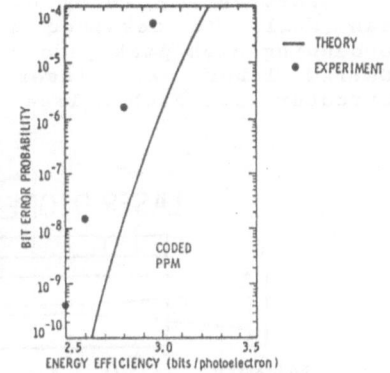

Fig. 5. Experiment   Fig. 6. Theory vs Exp.

On the left is a semiconductor laser diode operating at 0.83um wavelength. Its output beam is collimated and passed through 70-100 dB of neutral density filters (attenuator glass). The resulting weak signal is applied to a photomultiplier tube enclosed in a thermo-electric cooler housing. The photomultiplier tube is biased for the detection of single photon arrival events. The pulse position modulator and demodulator is located at the far right of the figure.

Figure 6 shows the comparison of the theory and the experimental results. The error probability is plotted as a function of the energy efficiency measured in bits/detected photon. The experimental results agree extremely well with the theory, and demonstrate that from 2.5 to 3.0 bits/detected photon performance was achieved depending on the acceptable value of error rate desired.

B. Laser Technology

Initial technology developments in the laser area concentrated on semiconductor lasers. Their small size, high power efficiency, rugged semiconductor construction, and the prospect of achieving the required average power levels, made them extremely attractive. However, deep-space communications, particularly where performance comparable to the previous section is desired, requires highly peaked signal intensities. Although the average power output levels may be only a few Watts (which has already been achieved in semiconductor lasers), the peak power requirements are in the 1-5 KW region. This is to take full

power-efficient use of the PPM modulation and to reduce
susceptibility to background light by smaller time-window
aperturing. Since semiconductor lasers usually suffer from
catastrophic facet damage at less than 10 times their rated
average power levels, semiconductor lasers, by themselves, were
considered inadequate.

Fortunately, many of the benefits of semiconductor lasers
can still be retained in other laser structures capable of
producing high peak powers. The one of most interest has been the
Nd:YAG laser with laser diodes used for pumping. The basic
structure for such a laser is shown in figure 7.

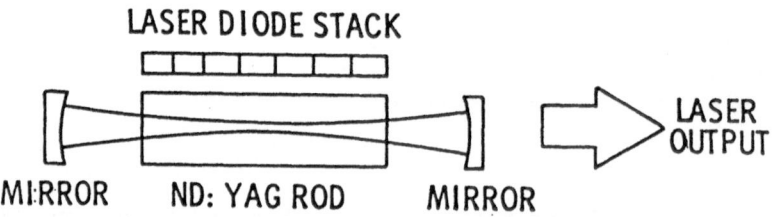

Fig. 7. Conventional Nd:YAG Architecture

Here, laser diodes, matched to the absorption band of the Nd:YAG
rod, are used to excite the rod from the side. Cavity mirrors are
placed at both ends of the rod to stimulate emission through the
center of the rod.

The primary disadvantage of such lasers was that the overall
wall-plug power efficiency was too low. Power efficiencies in the
0.1 to 0.5 % range were typical. However, it was noticed that the
region of the rod that was being pumped the hardest, the outside
circumference, was not the same region where the cavity mirrors
were trying to stimulate emission (in the center of the rod).
Thus, a new design using axial pumping, where the laser diode
energy is passed through an anti-reflection coating on the rod's
end, was developed. This resulted in a spatial matching of the
pump mode and the lasing mode in the rod. Additionally, the
diameters of the two matched modes were decreased so that the
power density, and hence the power conversion efficiency, was
increased. The resulting design produced an overall wall-plug
power conversion efficiency of 8.5% [Refs 8,9]. Since that time,
similar designs have produced up to 17% overall power efficiency
[Ref 10]. A picture of the 8.5% laser is shown in Figure 8.The
laser cavity was extremely robust, and could be assembled from
the individual components in about 30 minutes. Additionally, the
internal cavity gain was so high that simply inserting a
nonlinear crystal (KTP) into the cavity would produce second
harmonic radiation without either squelching the lasing or the
need to redesign the cavity to reduce losses.

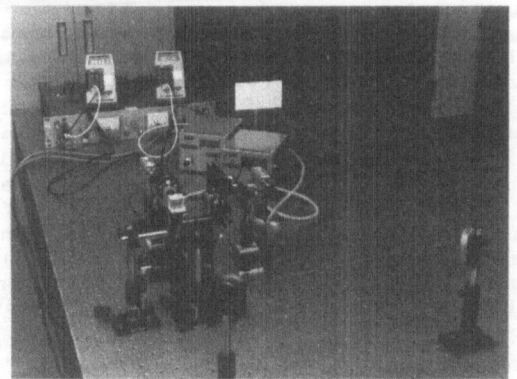

Fig. 8. 8.5% Efficient Nd:YAG Laser

   More recently, this basic laser concept has been packaged
into a custom laser module [Ref 11]. Figure 9 shows the resulting
assembly.

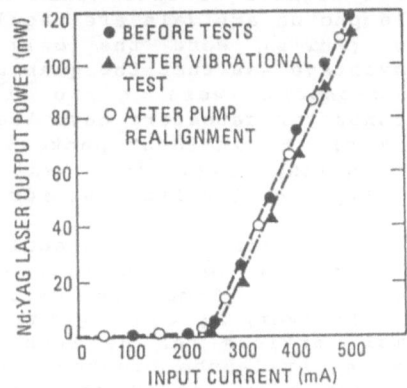

Fig. 9. Custom Nd:YAG Module   Fig. 10. Laser Test Results

It consists of two 10-stripe laser diode arrays used as pump
sources. These are combined in a polarizing beam splitter to
double the pump power. Collimation and apodization optics are
press-fit into the module in the regions between the two copper
laser diode array holders and the beam splitter. Following the
beam splitter is a focusing lens to focus the combined pump
energy into the Nd:YAG rod. The back of the rod, and the output
coupler mirror (far right) are coated to form the basic lasing
cavity. There are no operational adjustments on the unit, only
slotted holes in the laser diode heat sinks for initial alignment
of the pump beams during assembly. The remaining components were
cemented into place using space-qualified epoxy.

   Upon assembly, the unit was tested and produced 130 mW of
single spatial mode output power when using only two 200 mW laser

diode arrays as pumps. The overall power efficiency of the unit was about 7%. After testing, the module was subjected to three-axis vibrational shaking using the Space Shuttle launch vibrational spectrum, and then later thermal cycling typical of launch conditions. The results are shown in Figure 10. After shaking, the output power dropped by about 9%. Thermal cycling had no additional effect on performance. After completion of the environmental tests, the unit was reexamined. It was found that one of the laser diode heat-sink holders had shifted slightly during the shaking. By repositioning that diode and heat-sink, full power was restored from the unit. The conclusion was that with a little more care in the design of the heat-sink fastening, the unit would have survived all of the environmental test without degradation.

C. Detectors

Not only is it important to efficiently generate the optical power, it is also equally important to provide the utmost in efficiency when detecting it. The weakness of the received optical signals dictates that detectors capable of responding to single photon arrivals are required. In the past, photomultiplier tubes (PMT's) were the only such detectors. Their biggest disadvantage was that the quantum detection efficiencies of their photocathodes were quite low; typically less than 12%. Semiconductor detectors have the capability to convert as much as 60-75% of the incident photons to photoelectrons. Unfortunately their gains, even for avalanche photodiodes (APD's), have typically been insufficient for single photon detection.

Higher gains in APD's could be achieved with higher voltages if it weren't for the fact that thermal carriers trigger an avalanche and clamp the voltage at the avalanche saturation value. However, by cooling the device, the rate of generation of thermal carriers can be reduced sufficiently so that voltages above the avalanche breakdown can be sustained. Consider an APD which has been cooled and biased beyond the normal avalanche point, but which is not avalanche triggered. Upon the arrival of a photon, with high probability a free photoelectron will be generated. This carrier sees a tremendous potential field across the junction of the diode and accelerates rapidly to create a massive avalanche. This causes the applied voltage across the device to drop to the clamped avalanche breakdown level. Thus, for a single photon arrival, a detectable voltage drop is observed across the device. After a detection event, the voltage must be temporarily reduced (quenched) to reestablish the overbiased conditions. Because of the nonlinear nature in which these devices are used, the normal concept of excess noise associated with APD's has very little relevance.

Figure 11 shows the basic set-up for evaluating cooled APD's in this way [Ref 12].The output of a laser diode is attenuated, collimated, and applied to a cooled APD. Part of the optical beam is extracted by a beam splitter and detected by a PMT for calibration purposes.

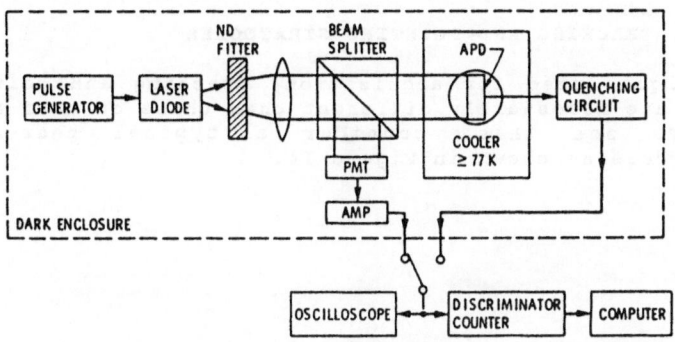

Fig. 11. Cooled APD Experimental Set-up

The APD under test is cooled and overbiased prior to detection, and quenched by a passive (load-line) quenching circuit after detection. Cooling of the APD can be as low as 77K, but it was found that operation around 200K was actually better. This is believed to be because of reduced carrier freeze-out at the higher temperature. A small Joule-Thompson cooler was used and the temperature was adjustable by controlling the vacuum level and nitrogen flow rate. Figure 12 shows the opened cooler with the APD mounted at the far left.

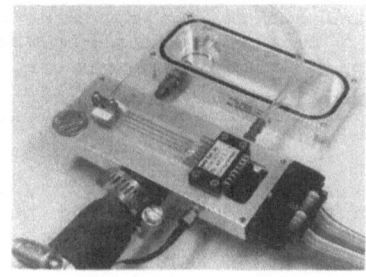

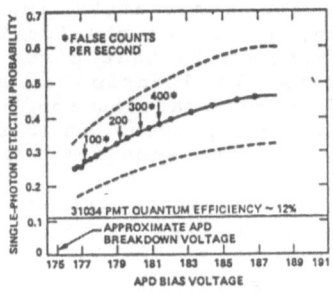

Fig. 12. APD with Cooler   Fig. 13. Detection Performance

The results of the detector evaluation are shown in Figure 13. Plotted here is the single photon detection probability as a function of the applied bias voltage. The value of the normal avalanche breakdown voltage is identified on the axis. The detection probability ranges from 25-40%, depending on the applied voltage. The number of residual dark counts/sec resulting from thermal carriers or other trapped-state release processes is shown parametrically on the curve. The dashed lines in the figure represent the expected uncertainty in the detection probability measurement.

D. ACQUISITION, TRACKING AND POINTING STRATEGIES

The techniques used for acquisition, tracking and pointing for deep-space are necessarily different than those of near-Earth crosslinks. To see this, consider a typical near-Earth acquisition process as shown in Figure 14.

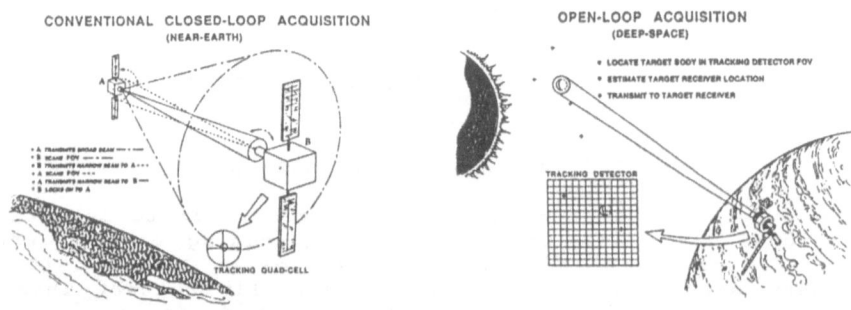

Fig. 14. Near Earth ACK   Fig. 15. Deep-Space ACK

Initially, neither spacecraft platform has adequate knowledge to precisely locate the other. The acquisition process is initiated from platform A by transmitting a broad beacon beam in the direction of Platform B. The width of the beacon is sufficient to cover A's uncertainty in B's location. Upon reception of the beacon, B can see the location of A and can return a narrow beam transmission in that direction. Once A receives the narrow beam, it knows the location of B with high precision and can replace its beacon transmission with a narrow one. To complete the acquisition process in this scenario requires beam propagation three times between the two platforms.

In deep-space communications, the one-way beam propagation time can be from several 10's of minutes to a couple of hours. Furthermore, by the time the beam reaches the other terminal, the original platform may no longer be in that location (eg a low altitude Earth-orbiting receiver may have a 90 minute orbit, spending 45 minutes of each 90 behind the Earth). Thus, an acquisition, tracking and pointing strategy which does not require two-way beam propagation is needed. Fortunately, at planetary distances, nature can provide the necessary spatial references in the form of the solar-illuminated planets themselves. Figure 15 illustrates this concept. Here a deep-space vehicle is coming out of occultation from behind a distant planet. As soon as there is a clear path to the Earth, the telescope on the deep-space vehicle is pointed toward the Earth. A two-dimensional detector array, large enough to cover the

attitude uncertainty of the spacecraft, is used at the focal
plane of the telescope to locate the Earth's image. By locking
onto that image with sufficient accuracy, adequate knowledge is
available to point and fire the return communication laser beam
at the Earth receiver. Since the solar-illuminated Earth image is
always available at the distant planet (except for relatively
infrequent singularity situations) the entire acquisition,
tracking and return beam pointing process can be accomplished in
a relatively short time period (likely under 30 seconds).

In the figure, the Earth image is depicted as covering
several 2-D detector pixels. Although there are realistic
situations where the Earth may appear as a point source, in most
situations the Earth will appear resolved. For example, from Mars
the Earth will appear from 4-20 pixels wide when viewed by a 10
cm telescope, depending on the Earth-Mars range. Since detector
pixels are usually sized to match the diffraction limit of the
corresponding telescope, then by reciprocity the transmitted beam
size, by the time it reaches the Earth, will correspond to the
portion of the Earth viewed by one of the detector pixels. For
this reason, it is necessary that the acquisition and tracking
system be able to locate an arbitrary point on the Earth (say its
geometric center) to a small fraction of a pixel. Additionally,
The Earth image will not appear as a smooth continuous image
superimposed on a detector grid (as shown in the figure), but
will be coarsely spatially quantized in both dimensions by the
resolution of the detector array. These difficulties necessitate
a more complex image correlation-type of acquisition and tracking
processor. Techniques which use image correlation and image edge-
tracking are being developed for these applications [Ref 13,14].

V. EARTH-END DESIGNS AND STUDIES

In addition to the spacecraft end, one must also have a good
understanding of the Earth reception structures and environment.
Of particular interest are such things as the cost and design of
the reception terminals, as well as the effects of the Earth's
atmosphere if the terminal is to be located on the ground. In
this section we discuss these topics.

A. Earth Reception Cost Model

It is envisioned that the first optical reception
capability, even if it is only for experimental purposes, will be
located on the ground. Additionally, it is of great benefit to
have some idea what the cost of ground-based facilities will be,
should it become desirable in the future to implement an entire
ground-based network. To get some insight into this area, data
from a large set of ground-based optical, infrared, and sub-
millimeter telescopes were collected and organized. Relationships
between cost, diameter and achievable blur circle (a measure of
the quality of the primary figure) were extracted. From this, a
cost model which depends on the diameter D, and the blur circle

field-of-view F, was formulated [Ref 15.]. Next, the communications performance achievable from a given telescope diameter and blur circle factor (which determines the susceptibility of the telescope to background noise) was calculated. Finally, the two models were combined into an overall cost verses communications advantage relationship and the result was minimized over diameter and blur circle. Communications advantage was measured against the current Deep Space Network's X-band capability and assumed nominal parameters for the optical spacecraft signal (30 cm telescope, 2 Watt laser, Saturn Distance). Figure 16 shows the output of this effort.

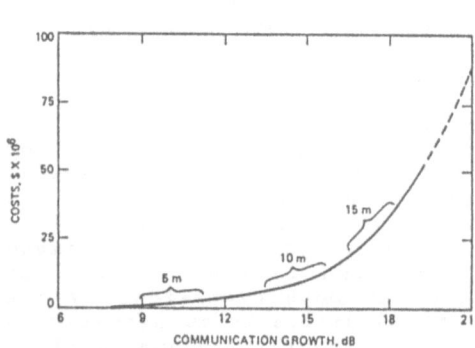

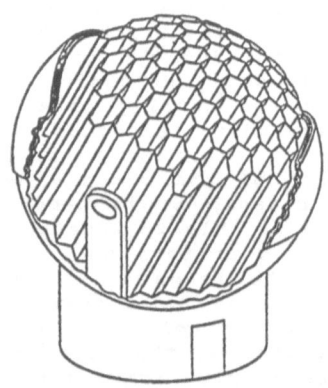

Fig. 16. Telescope Costs   Fig. 17. DSORA R+D STATION

We see that substantial communications performance can be realized without excessive cost if the telescope is up to 10 meters in diameter and with the appropriately chosen surface quality.

B. Ground-based R+D Station Design

One of the main ingredients of the current optical communications program is the design of a ground-based R+D station. Such a station would be used to support selected flight experiments as well as to demonstrate the technologies needed for future network deployments, whether on the ground or in space. Based on the above cost modeling efforts, and anticipated experimental needs, the design of a 10-meter reception facility was initiated. The primary aperture was taken as a segmented structure, consisting of light-weight composite materials. This was to keep the requirements down on the panel support structure and to facilitate extension of the technology into space in the future. One of the unique requirements on the facility was that it be usable during the day since many planetary spacecraft must utilize daytime tracking. Furthermore, tracking at small Sun-Earth-probe (SEP) angles is occasionally required. This must be accomplished without permitting direct sunlight to strike the primary surface since intolerable background light scattering will result. The structure must also be sufficiently compact to reduce wind loading and the entire structure must be of affordable cost.

Figure 17 shows a sketch of the current design [Ref 16]. It consists of a segmented primary of 5 segmentation rings. The panels are expected to be made of carbon-carbon, or similar composite material. To permit small SEP viewing, an integral sunshade has been included. The basic concept is similar to a bundle of soda straws. If each straw has a long length-to-diameter ratio, then small SEP viewing can be achieved even though the absolute length of the sunshade is comparable to the telescope diameter.

For background noise immunity, the reception station is expected to us a Fraunhofer filter. A Fraunhofer line is a spectral region of the Solar spectrum where very little light is produced. It is a result of molecular absorption in the plasma above the Sun's photosphere. A Fraunhofer filter is a very narrow filter which is tuned to one of these Fraunhofer lines. When viewed through a Fraunhofer filter, the Sun, or reflected sunlight from a planet, appear much darker.

Designs and construction planning are currently in progress for this R+D station. It is expected that the facility construction can be completed by the year 2000.

C. Atmospheric Visibility Characterization

One of the important variables in ground-based reception is the impact of the Earth's atmosphere, most importantly the cloud attenuation. This is important not only for quantitatively predicting the performance of a ground-based station, but also for quantifying the benefit that would be afforded by going to space later on.

Typically, the southwestern US has cloud-free visibility more than 60% of the time [Ref 17]. Although good for making first-cut estimates of performance, an average visibility model is insufficient for supporting a planetary flight project. A flight project manager would typically like to know the answers to such questions as "When my spacecraft reaches its destination in N years, what are the chances my data will be lost due to weather outages? How valid is the assumption of independent weather cells? What are the statistics for the weather fade outages and the fade depths? What seasonal dependencies do these quantities have?", and many more. To gain insight into the answers to these kinds of questions, an activity called the Autonomous Visibility Monitoring program was started [Ref 18-20]. The program has purchased a set of three computer-controlled and autonomously operated visibility monitoring telescopes, along with roll-off roof domes, and meteorological sensors to determine when the weather is safe enough for viewing. The telescopes will be used to track natural light sources (like planets, stars, or even the blue sky) both at night and during the day. Simultaneous observation of the same sources will be accomplished from three sites spread over the southwestern U.S. Data from the sites will be relayed back to a central computer at JPL via telephone modem where analysis will take place.

The first of the observatories has been deployed at Table Mountain Observatory in California (see Figure 18). The second and third observatories will be deployed soon at Mt Lemmon, AZ and Sacramento Peak, NM respectively.

Fig. 18. AVM Observatory at Table Mountain

VI OPTICAL COMMUNICATIONS LONG RANGE PLAN

The activities of JPL's deep-space optical communications program have all been coordinated around an overall long range development plan. The plan was originally developed in 1984 and has been maintained and updated since then [Ref. 21]. The plan is shown in Figure 19.

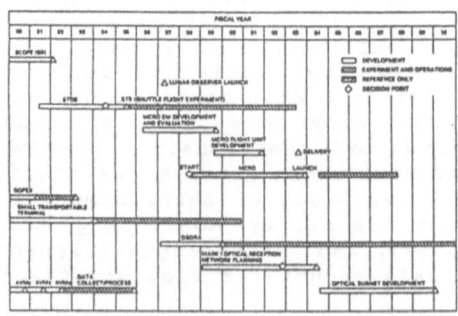

Fig. 19. Long-Range Plan

The upper portion of the plan applies to the space-end of the link whereas the lower portion refers to the ground or Earth-vicinity end. The space-end activities progress through the development of a Space Technology Demonstration System (STDS) which is expected to be a Space Shuttle-to-ground demonstration. This will be followed by a full-up mission enhancement experiment on the proposed Mars Communication Relay Orbiter (MCRO). The early ground-end activities include the deployment of the AVM observatories, and an uplink optical demonstration with the Galileo spacecraft (GOPEX). The GOPEX demonstration will use the Galileo imaging camera to receive optical pulses transmitted from a telescope at Table Mountain Observatory. Additionally, a small ground terminal for supporting future (late 1990's) space-to-ground demonstrations will be developed. This will be a precursor to a Deep-Space Optical Reception Antenna (DSORA) which will serve as a long-term R+D optical ground station. Knowledge obtained from these activities will be factored into the design and planning for the eventual deep-space optical reception network.

VII. CONCLUSIONS

In this paper the deep-space optical communications program has been described. The motivation for the use of optical frequencies was discussed and quantified. This was followed by a presentation of the various systems architectures. Technologies relevant to the spacecraft, and the Earth reception end, were covered. The design and cost modeling for a ground-based R+D reception station were discussed, as well as a program to characterize the visibility function of the atmosphere. The long-range development plan was then shown.

The mechanisms for overcoming the remaining foreseen challenges are in place and there appear to be no insurmountable barriers standing in the way. It is predicted that deep-space missions will conduct flight experiments of the technology in the late 1990's, and that operational use on future missions will occur shortly thereafter.

VIII. ACKNOWLEDGEMENTS

The author would like to express his appreciation to the members of the Optical Communications Group at the Jet Propulsion Laboratory who individually, as well as collectively, contributed to the research, development and planning that is reported in this paper. The research described in this paper was carried out by the Jet Propulsion Laboratory, California Institute of Technology, under contract with the National Aeronautics and Space Administration.

IX. REFERENCES

[1]   Shaik, K. S., "A Preliminary Weather Model for Optical
      Communications Through the Atmosphere," TDA Progress Report
      42-95, Jet Propulsion Laboratory, Pasadena, Calif.,pp.212-
      218,November 1988.

[2]   Levitin, L.B.,"Photon Channels With Small Occupation
      Numbers," Problemy Predachi Informatssi, Vol 2,No.2,pp.60-
      68,1966.

[3]   Pierce, J.R.,"Optical Channels: Practical Limits With Photon
      Counting," IEEE Trans. Comm, Vol.COM-26,NO.12,pp.1819-
      1821,Dec.1978.

[4]   McEliece, R.M.,"Practical Codes for Photon Communications,"
      IEEE Trans. Information Th.,Vol.IT-27,1981.

[5]   Lesh, J.R.,"Optical Communications Research to Demonstrate
      2.5 Bits/Detected Photon," IEEE Communications Society
      Magazine,pp.35-37,November,1982.

[6]   Lesh, J.R.,et al,"2.5 Bit/Detected Photon Demonstration
      Program:Description, Analysis, and Phase I Results," TDA
      Progress Report 42-66, Jet Propulsion Laboratory, Pasadena,
      Calif.,pp.115-132,November 1981.

[7]   Katz, J.,"2.5 Bit/Detected Photon Demonstration Program:
      Phase II and III Experimental Results," TDA Progress Report
      42-70, Jet Propulsion Laboratory, Pasadena, Calif.,pp.95-
      104,July 1982.

[8]   Sipes, D.L.,"Highly Efficient Neodymium:Yttrium Aluminum
      Garnet Laser End Pumped by a Semiconductor Laser Array,"
      Appl.Phys.Lett.,44(2),pp.74-76,July 15,1985.

[9]   Sipes, D.L.,"Method and Apparatus for Efficient Operation of
      Optically Pumped Laser," U.S.Patent 4,710,940 December 1,
      1987.

[10] Skrlac, W.J. and H.P.Kortz,"High Power, High Frequency
      Neodymium Laser End Pumped by a Single-Stripe Laser Diode,"
      Proceedings of LEOS 88, Santa Clara, Ca,paper EL1.2,
      November 1988.

[11] Hemmati, H. and J.R.Lesh,"Reliability Testing of a Diode-
      Laser-Pumped Nd:YAG Laser and a Set of Diode-Laser Arrays,"
      Proceedings of SPIE OE LASE 89, paper 1059-22,January 1989.

[12] Robinson, D.L. and B.D.Metscher,"Photon Detection with
      Cooled Avalanche Photodiodes," Appl.Phys.Lett.
      51(19),pp.1493-1494,November 9, 1987.

[13] Win, M.Z.,"Estimation and Tracking for Deep-Space Optical
      Communications," Proceedings of SPIE OE LASE 89, paper 1059-
      12, January 1989.

[14] Chen, C.C. and M.Z.Win,"Effects of Earth Albedo Variation on
      the Spatial Acquisition Subsystem of a Planetary
      Spacecraft," Proceedings of SPIE OE LASE 89, paper 1059-9,
      January 1989.

[15] Robinson, D.L. and J.R.Lesh,"A Cost-Performance Model for
      Ground-Based Optical Communications Receiving Telescopes,"
      Proceedings of SPIE OE LASE 89, paper 756-20, pp.130-134,
      January 1987.

[16] Kerr, E.L.,"Architectural Design of a Deep-Space Optical Reception Antenna," submitted for publication in Optical Engineering, January 1989.

[17] Wylie, D.P.,"Cloud Cover Statistics From GOES/VAS Satellite," Proceedings of SPIE OE LASE 88, paper 874-34, January 1988.

[18] Cowles, K.A.,"A Visibility Characterization Program for Optical Communications Through the Atmosphere," TDA Progress Report 42-97, Jet Propulsion Laboratory, Pasadena, Calif., May 15,1989.

[19] Erickson, D.M. and K.A.Cowles,"Options for Daytime Monitoring of Atmospheric Visibility in Optical Communications," TDA Progress Report 42-97, Jet Propulsion Laboratory, Pasadena, Calif., May 15, 1989.

[20] Cowles, K.A.,"Site Selection Criteria for the Optical Atmospheric Visibility Monitoring Telescopes," TDA Progress Reports 42-97, Jet Propulsion Laboratory, Pasadena, Calif., May 15,1989.

[21] Lesh, J. R., L. J. Deutsch, and W. J. Weber, "A Plan for the Development and Demonstration of Optical Communications for Deep-Space," TDA Progress Report 42-103, Jet Propulsion Laboratory, Pasadena, CA Nov. 15, 1990.

DLR Experimental Systems for Free Space Optical Communications

J. Franz,* Ch. Rapp and B. Wandernoth

German Aerospace Research Establishment (DLR)

Institute for Communications Technology

D–8031 Oberpfaffenhofen

Abstract

Two coherent optical transmission systems, both transmitting 565 Mbit/s at 1064 nm are presented. These systems have been designed for experimental studies in the laboratory and in a stationary free space optical link. The aim is to demonstrate the suitability of such high–bitrate coherent optical data transmission schemes for future space applications. The prototype breadboard system uses PSK modulation with homodyne reception. A new synchronization technique made it feasible to build up a low–complexity receiver allowing a power efficient transmission 3.5 dB above the shot noise limit. The second system is a compact and robust DPSK system with heterodyne detection. This system is embedded in a new stationary test facility, which shall allow free space communication experiments between two buildings (760 m).

1 Introduction

Free space optical communication for satellite and deep space applications is a challenging aim for the next decades. Advances in the development of fast electronic circuits and high quality optical components, such as lasers and optical modulators, allowed the realization of very high sensitive and high speed coherent optical communication systems in the past. For future space applications, such systems will offer many advantages over the traditional microwave transmission schemes, such as reduced receiving and transmitting antenna size, higher transmission capacity, avoidance of interference problems in the overcrowded microwave fequency spectrum and many others.

---

*Dr. Franz is now professor for Optical Communication at the Polytechnical University Düsseldorf

Much work has been already done in this field [1,2,3,4,5,6]. Since 1989, also the DLR institute for communications technologie has spread its activities in digital communication systems up to the optical frequencies. In the following, the DLR experimental systems for free space optical communications will be presented.

In section 2 a breadboard PSK homodyne system with a special technique for carrier synchronization will be presented. This system allows a power efficient operation 3.5 dB above the shot noise limit with 565 Mbit/s. For testing and demonstration of the components and the systems under more realistic conditions a stationary 760 m free space link between two buildings has been established. This testbed, which is presently equipped with a 2–DPSK optical heterodyne system, will be presented in section 3.

2　The Optical Homodyne System

Free space optical communication between satellites and especially for deep space applications require the most sensitive transmission scheme as possible. As is shown in theory, the best performance in this sense offers an optical PSK system with homodyne detection [7,8]. In such a system some kind of an optical PLL is required, that allows the local oscillator laser (LO) to be tuned to exactly the same frequency and phase as the transmitted carrier. Normally, the phase locking of the LO in an optical homodyne system is one of the most critical problems in the realization. The DLR breadboard PSK homodyne experimental system, which uses a new variant of a carrier synchronization technique is presented in the following.

2.1　PSK Homodyne Receiver Synchronization Using Synchronization Bits

Phase control of the local oscillator in an optical PSK homodyne system is usually performed by means of the Costas loop technique [9]–[15] or the pilot carrier technique [16]–[23] In spite of higher complexity and the need of an optical hybrid the Costas loop is often preferred to the simple pilot carrier technique because AC coupling is possible and the PLL is independent of the data signal [9,10,24,25]. In the following new approach, the advantages of the Costas Loop are combined with the simplicity of the pilot carrier receiver.

Proceeding from the Costas Loop design, the principle of the synchronization method is shown in figure 1 and 2. Instead of feeding a fraction α of the received light into the quadrature arm continuously as done in a Costas Loop receiver (fig. 1), the inphase arm and quadrature arm are used alternately (fig. 2). It can easily shown, that if the quadrature arm is used for a portion α of time, the properties of this receiver are the same as compared to a Costas Loop receiver with splitting ratio α. The phase error

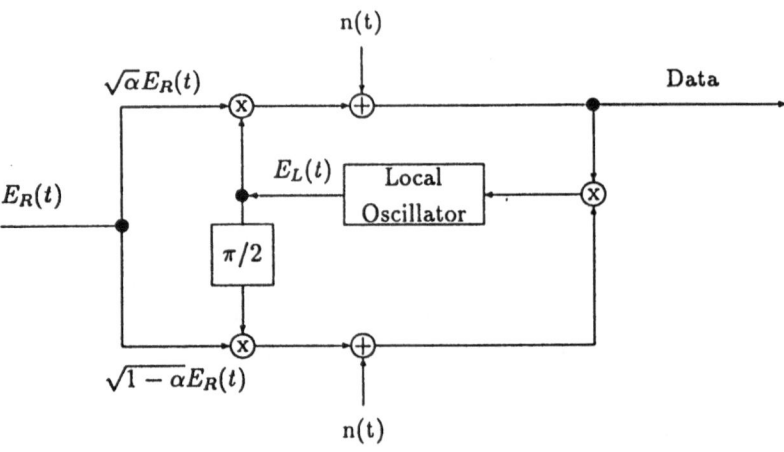

Figure 1: Principle of the PSK homodyne receiver with a conventional Costas loop.

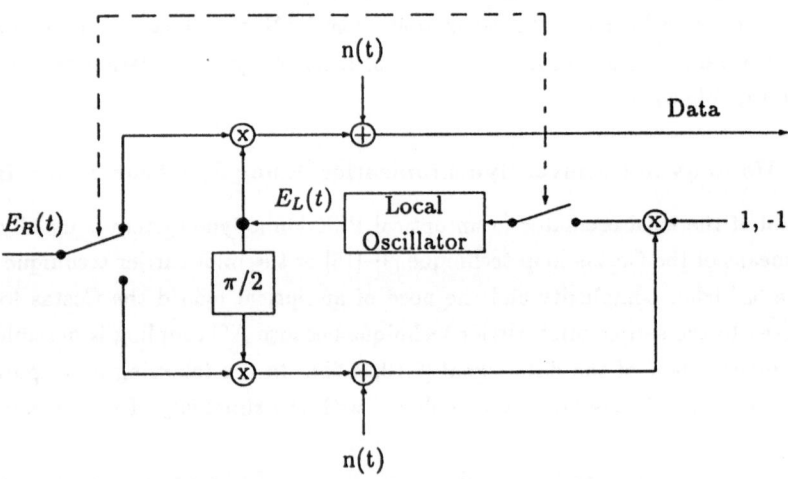

Figure 2: Principle of the sync–bit synchronization technique derived from the Costas loop design, where inphase and quadrature arms are used alternately.

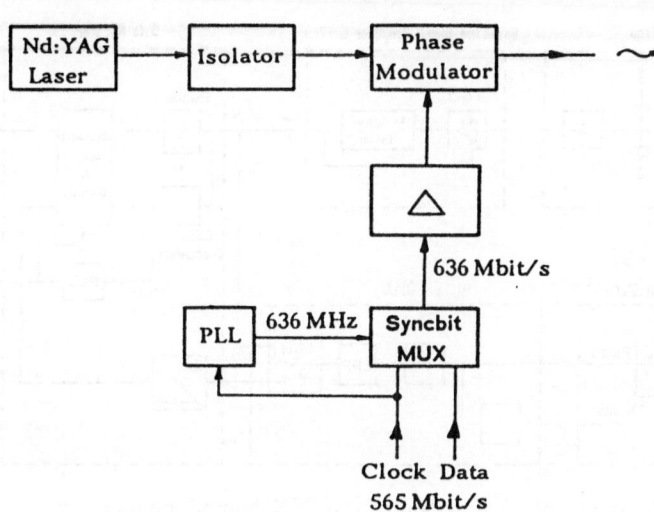

Figure 3: Block diagram of the PSK transmitter.

signal is obtained by sampling the quadrature signal and multiplying it with the polarity of the bit sent at this time.

The switching of the received light power to the quadrature arm and the generation of the phase error signal could be done at every bit for the portion of time αT_B ($T_B =$ duration of one bit) or every $1/\alpha$ bit for the time T_B. Which technique is applied depends on the laser linewidth–to–bitrate ratio. If entire bits are used for synchronization as shown in figure 1, these bits are called synchronization bits.

As can be seen in figure 2, inphase and quadrature arms are never used at the same time. Therefore, with regard to the practical realization, only one optical frontend and one signal path is sufficient. Switching between the inphase and quadrature functions of this signal path could be simply obtained by shifting the phase of the local oscillator by 90 degrees at the right moments, but there will be still the need for a phase modulator in the receiver. The most elegant method with lowest complexity provides the 90 degree switching in advance in the transmitter, where a phase modulator is present at all. This technique has been realized in a experimental setup in the laboratory.

2.2 Homodyne Experimental Setup

The principle block diagram of the homodyne transmitter is shown in figure 3. The

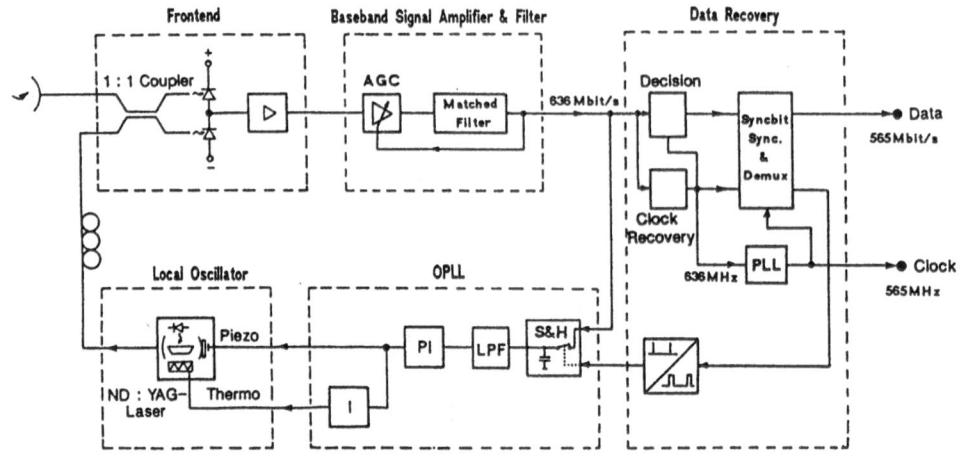

Figure 4: Block diagram of the PSK homodyne receiver.

optical path contains a diode pumped Nd:YAG laser with wavelength $\lambda = 1064$ nm and $P_s = 45$ mW output power. The laser is screened against backreflections by the optical isolator. The external modulator is a travelling wave LiNbO$_3$ phase modulator. Prior to phase modulation ("1" $= +90^0$, "0" $= -90^0$), the incoming data stream is supplied with the synchronization information by insertion of one synchronization bit (0^0) after every eighth bit. An example for the resulting modulating signal with the sync–bits and the expanded symbol rate of 635.625 Mbit/s is shown in the upper trace in figure 5.

The PSK homodyne receiver is shown in figure 4. It consists of a simple balanced frontend with transimpedance amplifier. The level of the shot noise is about 8 dB higher than the thermal noise (LO power $P_1 = 4$ mW). The signal then passes to an amplifier with automatic gain control (AGC) and a matched filter. Since the pulse shape of the transmitted signal is rectangular, the detected signal after the matched filter has a triangular form, as shown in the middle trace of figure 5. The following clock recovery regains the 635.625 MHz clock even when an optical phase lock is not yet achieved. This is required for the following digital logic, which finds the synchronization bits, to sample the analog signal at the right moment to obtain the phase error signal. The frequency and phase of the local oscillator are controlled by temperature and a piezo crystal. The synchronization bits still contained in the detected signal after the matched filter (see fig. 5) are removed by means of a demultiplexer, and the original data with 565 Mbit/s is regained (fig. 5,lower trace).

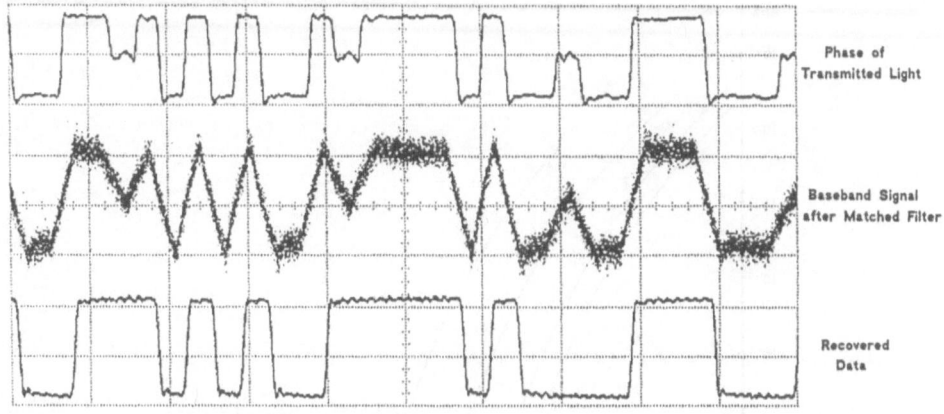

Figure 5: Exemplary waveforms in the realized transmission system (5 ns/division): Modulating signal (phase of the transmitted light), baseband signal after matched filter and recovered data stream.

2.3 Performance Results

The measurement setup for the performance evaluation of the homodyne system is shown in a short form in figure 6. The transmitted digital signal is a pseudo–random sequence of length $L_p = 2^{10} - 1$. The signal is transmitted over a short free space distance on the optical bench and attenuated by defocussing the telescopes. A 3 dB coupler splits the incoming light so that the optical power meter indicates the power of the light directly at the input of the receiver. This method is easier and more accurate than deriving the input

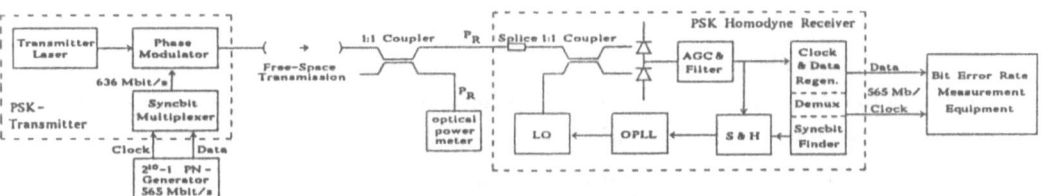

Figure 6: Measurement setup for the PSK system.

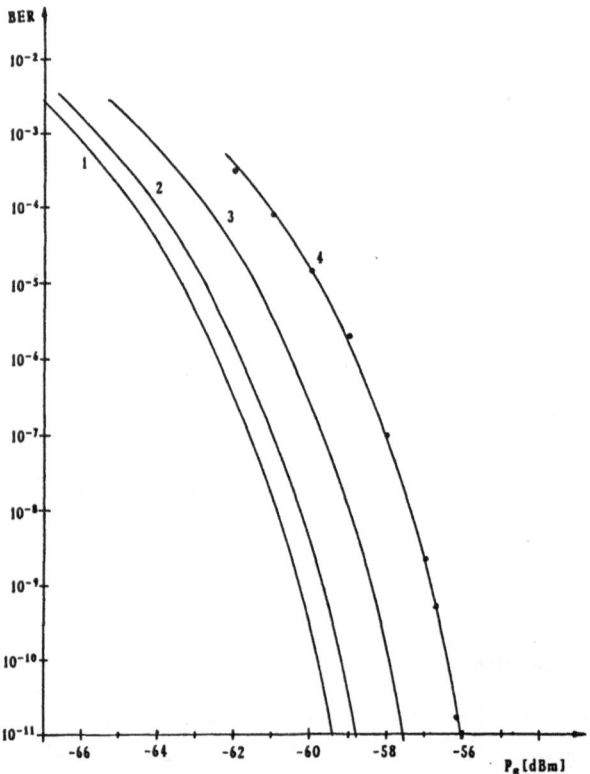

Figure 7: Bit error rate against optical input power: (1) ideal system, shot noise limited, (2) system performance including loss due to synchronization, (3) with additional loss due to quantum efficiency and (4) experimental result.

power from the IF current level. Measurement accuracy depends only on the asymmetry of the coupler, loss of the splice and accuracy of the power meter. The asymmetry of the coupler can easily measured and has been taken into account; the splice has been considered to be a part of the receiver and the accuracy of the power meter is better than 10% in the nanowatt range. Therefore the maximum error is approximately 0.4 dB. The bit error rate (BER) has been measured as a function of the received optical power P_R in the receiver and the results are shown in figure 7.

Curve 1 represents the ideal 565 Mbit/s PSK homodyne receiver assuming photodetectors with 100% quantum efficiency and with no loss due to synchronization. As compared to this ideal curve, the realized receiver (curve 4) requires 3.5 dB more light power.

The degradations in figure 7 can be explained as follows:

Wavelength	$\lambda = 1064$ nm (Nd:YAG)
Information bit rate	$f_B = 565$ Mbit/s
Channel bit rate	$f_C = 635.625$ Mbit/s
Bit error rate	$BER = 10^{-9}$
Data (pseudo random)	$L_p = 2^{10} - 1$

Received light power required for an ideal system [8] (shot noise limit)	-60.3 dBm

Losses:

Increase of channel bit rate by 9/8	0.5 dB
Quantum efficiency of photodiode ($\eta = 0.75$)	1.3 dB
Other Implementation losses	1.7 dB

Received light power required in the realized system	Σ	-56.8 dBm

Table 1: Characteristics of the realized PSK homodyne system.

• The sync–bit carrier recovery technique described above causes a 0.5 dB sensitivity penalty (curve 2), because after every eighth data bit, a non information carrying sync–bit is inserted. This causes an increase of the channel bitrate by the factor 9/8 which results in a degradation of $(10 \log(9/8) = 0.5)$ dB, due to the increased receiver filter noise bandwidth.

• The non–ideal quantum efficiency of the photodiodes $\eta \approx 75\%$ causes further $(-10 \log 0.75 \approx 1.3)$ dB degradation (curve 3).

• The residual loss of 1.7 dB (curve 4) represents the effects of the thermal noise of the first amplifier and non–ideal baseband filtering and data clock recovery.

The performance and the technical characteristics of the PSK homodyne system are also summarized in table 1. The measured performance of this realized PSK homodyne system can be characterized alternatively by the need of 20 photons per bit. This is, to the best of our knowledge, the best sensitivity obtained to date with a coherent optical communication system at 565 Mbit/s.

3 The Free Space Testbed

3.1 Objective

In order to have a testbed for free space experiments the DLR institute of communications has built up a stationary test facility allowing free space optical communication between two buildings. The aim is to have a flexible facility for experiments under nearly real conditions for testing systems and components for future coherent optical space transmission schemes. The possible experiments are manifold like e.g.

- gathering experiences with different telescopes under nearly farfield reception conditions,

- testing of high power lasers or laser amplifiers used in future optical space transmission systems,

- on-earth experiments concerning the pointing, acquisition and tracking (PAT) problems,

- investigation of athmospheric effects on coherent optical transmission schemes and

- channel measurements of the athmospheric link.

Due to the open concept, also experiments with components and systems of external partners could be foreseen.

3.2 The Arrangement of the Testbed

A shelter with all optical and electronical equipment for the transmitter and the receiver is positioned on the flat roof of the institute building (5 floors). The shelter has two openings for the optical input and output beams. The optical link is established by choice with a mirror or a retroreflector, which is positioned on the roof of a second building with equal height. For a principal view of this facilities see figure 8. The shelter has been installed in the late 1991 and after the installation of the optical bench in the shelter, the first experiments using a coherent 565 Mbit/s DPSK heterodyne system [26] have been be started in April 92.

3.3 The Optical DPSK System for Free Space Experiments

The differential phase shift keying (DPSK) modulation scheme has been chosen because the obtainable receiver sensitivity is nearly as high (3dB difference) as with the theoretical best receiver (the PSK homodyne receiver), but no phase stabilization of the local

Front View of the Shelter

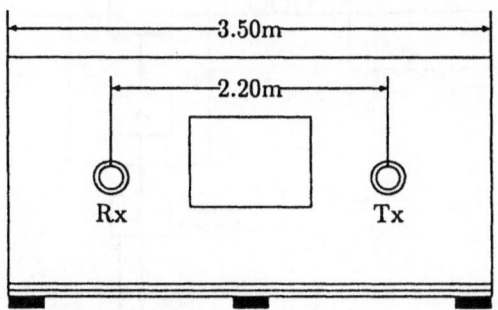

- trussed beton floor
- trussed front panel
- 2 separate telescopes (Tx and Rx)

Siteplan of the Optical Free Space Link(Top View)

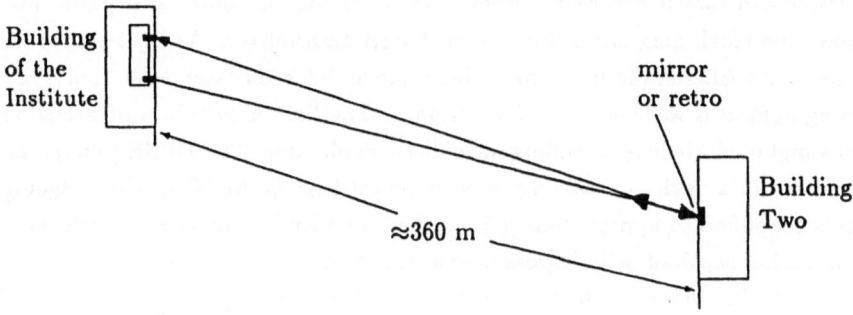

Figure 8: The principal arrangements of the shelter and the optical link.

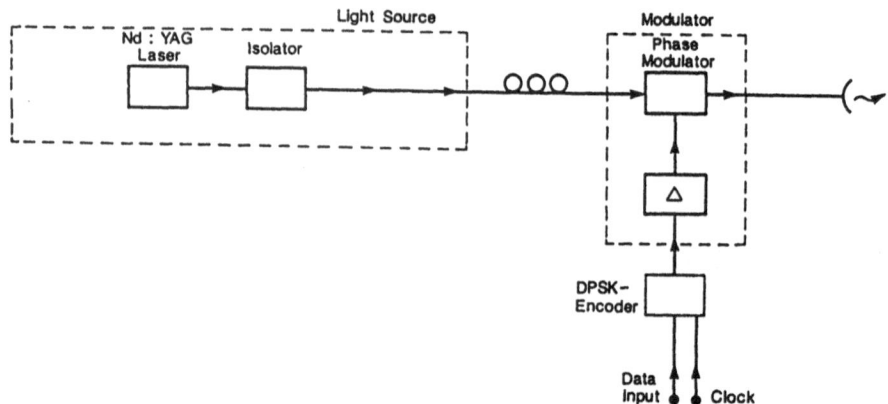

Figure 9: Block diagram of the DPSK transmitter.

oscillator is required. The realized system runs with a bitrate of 565Mbit/s. A compact transmitter and receiver design and the need of only one single 12 Volt power supply made this simple and robust coherent system suitable for the first outdoor experiments. Figure 9 shows the block diagram of the optical DPSK transmitter. As light source we used a commercial available diode-pumped single mode Nd:YAG laser with 50mW cw power emitting light at a wavelength of $\lambda = 1064$nm. The light is led via a polarization controller to a pigtailed LiNbO$_3$ travelling wave phase modulator. The DPSK differential encoder is realized in a device containing three standard logic GaAs IC's. The outgoing data stream is amplified to approximately 7 V$_{pp}$ and amplitude controlled to achieve a constant modulation depth of ±90 degrees over a wide temperature range.

Figure 10 shows the block diagram of the receiver. The incoming light is mixed with the polarization adapted light of a 4 mW MISER in a balanced frontend consisting of a symmetrical (k=0.47) 2x2 coupler and two InGaAs Photodiodes ($R_{eff} \approx 0.5$ A/W). The signal current is amplified by a transimpedance amplifier with an average noise power density of 14pA/$\sqrt{\text{Hz}}$. The intermediate frequency (IF) is only twice the data rate, i.e. 1130 MHz. After bandpass-filtering and amplification, the conversion into the baseband is done by a delay-line demodulator. The data signal is low-pass filtered in a gaussian filter with normalized cutoff frequency $f_{3dB}T_B = 0.79$ and then led to the clock and data recovery unit. This device is a commercially available hybrid circuit using analog and digital GaAs chips, which has been modified for better threshold adjusting due to the

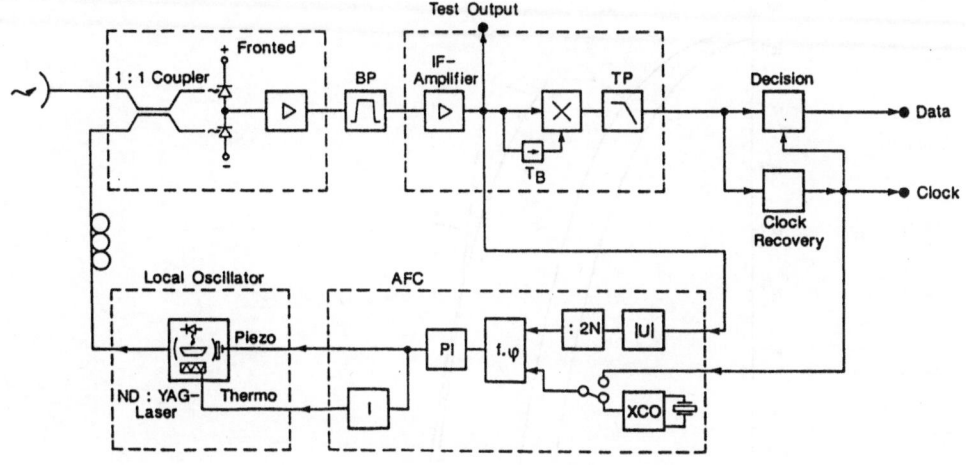

Figure 10: Block diagram of the DPSK heterodyne receiver.

demands of the DPSK scheme. Data and clock signals are available in ECL level.

Although there is no need to control the phase of the local oscillator in a DPSK heterodyne receiver, the intermediate frequency is to be stabilized by an AFC. In the AFC of this receiver the IF signal is squared and then bandpass-filtered and limited. Latter operations are required because of the amplitude distortions due to low IF and IF bandpass filtering. The so obtained 2.26 GHz signal is divided by 4 and compared with the recovered clock or a crystal oscillator signal. Frequency control of the local laser is achieved by using a piezo element and a temperature control unit. With the piezo element, fast but low–range frequency control is possible (±10MHz, speed up to 100 kHz), whereas the temperature control is used to compensate slow drifts by integrating the piezo signal. So the tuning range is only limited by the distance of mode hops of the local laser. Due to the filtering of the squared IF signal, the range in which the AFC is able to recognize a signal is only ±10 MHz. Therefore, the AFC has been supplied with a special device for frequency aquisition. If the IF deviation is larger than 10 MHz, no signal can be detected by the AFC and an oscillator at the frequency divider's input with approximately 2.5 GHz simulates an IF, which is too high. When the IF comes into the detection range, this oscillator signal is overdriven and frequency control begins. So care has to be taken, that when switching on the system, frequencies of transmitter laser and LO should be adjusted to a high intermediate frequency.

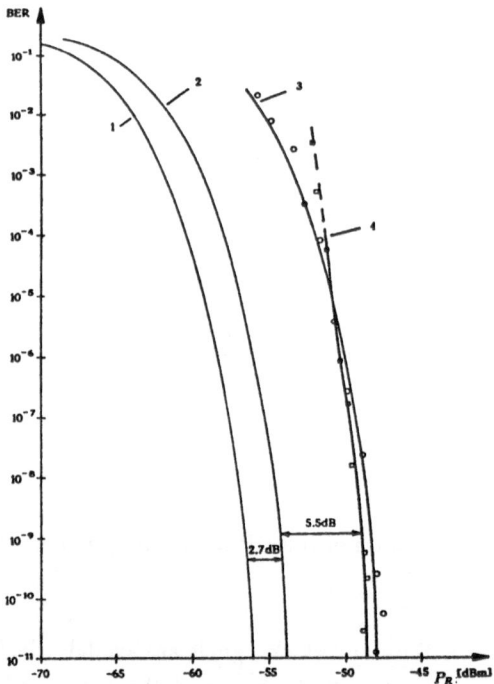

Figure 11: Bit error rate versus optical input power (1) ideal, (2) best realizable regarding the quantum efficiency of the diodes and the gaussian filter, (3) the realized system with hardwired clock and (4) with using the data clock recovery unit.

3.4 DPSK System Performance

The sensitivity of this receiver has been measured by using an IF power measurement technique [26]. Figure 11 shows the BER of the experimental DPSK system versus the optical input power. The ideal receiver at the shot noise limit is represented by curve 1, in curve 2 non-ideal quantum efficiencies of the photodiodes and the non-ideal baseband filter are taken into account. Curve 3 shows the sensitivity of the receiver described here without the data clock recovery unit, that means, that the data signal was sampled and decided in the receiver of the BER measurement equipment using the clock from the transmitter. The loss compared to curve 2 is mainly due to the high thermal noise of the transimpedance amplifier. Here some improvements will be made in the future. Other significant losses come from imperfect light coupling to the photodiodes, Finally, curve 4 shows the behavior of the BER of the complete system, i.e. with data clock recovery unit. One can see, that this device improves the sensitivity at low bit error rates,

Wavelength	$\lambda = 1064$ nm (Nd:YAG)
Bit rate	$f_B = 565$ Mbit/s
Bit error rate	$BER = 10^{-9}$
Data (pseudo random)	$L_p = 2^{23} - 1$

Received light power required for an ideal system [8](shot noise limit)	-56.3 dBm

Losses:

Quantum efficiency of photodiode ($\eta \approx 0.75$)	1.3 dB
Non–ideal baseband filter (Gaussian)	1.4 dB
Noise and and frequency response of first amplifier	2.5 dB
Non–ideal modulation (driver for modulator)	1.5 dB
Other implementation losses (especially demodulator)	1.6 dB

Received light power required in the realized system	Σ	-48.0 dBm

Table 2: Characteristics of the realized DPSK heterodyne system.

but the transmission nearly breaks down at BER's higher than 10^{-4}, because increasing phase jitter of the clock sometimes causes a synchronization loss of the BER measuring equipment producing more than 10 errors at such a moment.

For all measurements we used a pseudo-noise pattern with a length of $2^{23} - 1$ and a bit rate of 565 Mbit/s. No error rate floor up to $BER = 10^{-14}$ has been observed. The measured sensitivity of $P_R = -48$ dBm of this system at $BER = 10^{-9}$ corresponds to a detection of 150 photons per bit. The characteristics of this DPSK system have been summarized in table 2. The AFC with automatic frequency aquisition allows the system to be switched on by only one switch. After a few seconds the system runs stable and transmits data with a bit error rate corresponding to the received light power. In a long term experiment over several weeks in the laboratory we could prove the stability of the system even under temperature variations caused by sunlight during daytime. This qualified the DPSK system now to be placed into the free space testbed described above. Besides data transmission for demonstration purposes, also a digitized video signal can now be transmitted.

3.5   Free Space Experiments

First experiments in April 92 have shown the arising of many problems, which have already been expected, e.g.

- the problem of the mechanical stability of the whole system, including the building on which it is installed,

- mechanical and thermal stability of the fitting and positioning elements of the telescopes and the coupling of the received light into the monomode fiber coupler,

- vibrations of the mirror,

- optical quality of the mirror respectively the retro,

- coherence losses and polarization disturbances due to athmospheric effects.

These effects are now going to be investigated, looking forward to report on the first experimental results in the future.

4   Conclusions

We have presented the realization of two experimental systems for coherent optical space communications. The very high sensitive PSK homodyne system has been built up as a breadboard system using a new method for carrier synchronization. This technique is similar to the Costas loop design, but requires less critical optical components and fewer analog RF components. By using synchronization bits, optical and analog components have been replaced by digital circuits, which simplifies monolithic integration and therefore makes it useful in critical environments, like in future space applications.

The less critical, but also less sensitive DPSK heterodyne system has been realized in a compact and robust form, so that it could be installed in the new test facility for free space communications, described above. This system will be a cheap and convenient playground for realistic free space experiments in the forefield of future satellite experiments.

References

[1] *Proc. SPIE: Free Space Laser Communication Technologies*, vol. 885, 1988.

[2] *Proc. SPIE: High Data Rate Athmospheric and Space Communications*, vol. 996, 1988.

[3] *Proc. SPIE: Optical Space Communication I*, vol. 1131, 1989.

[4] *Proc. SPIE: Free Space Laser Communication Technologies II*, vol. 1218, 1990.

[5] *Proc. SPIE: Free Space Laser Communication Technologies III*, vol. 1417, 1991.

[6] *Proc. SPIE: Optical Space Communication II*, vol. 1522, 1991.

[7] T. Okoshi and K. Kikuchi. *Coherent Optical Fiber Communications*. KTK Scientific Publishers, Tokyo, 1988.

[8] J. Franz. *Optische Übertragungssysteme mit Überlagerungsempfang*. Springer-Verlag, Berlin Heidelberg New York, 1988.

[9] F. M. Gardner. *Phaselock Techniques*. John Wiley, New York, 2 edition, 1979.

[10] E. Gottwald et al. 2.5 GBit/s PSK homodyne system with nonlinear phase–locked loop. In *Proc. 16th European Conf. on Optical Communications*, pages 331–334, Amsterdam, 1990.

[11] T. G. Hodgkinson. Costas loop analysis for coherent optical receivers. *Electron. Lett.*, 22:394–396, 1986.

[12] T. D. Stephens and G. Nicholson. Optical homodyne receiver with a six–port fibre coupler. *Electron. Lett.*, 23:1106–1108, 1987.

[13] A. Schöpflin et al. PSK optical homodyne system with nonlinear phase–locked loop. *Electron. Lett.*, 26:395–396, 1990.

[14] S. Norimatsu and K. Iwashita. 10 Gbit/s optical PSK homodyne transmission experiments using external cavity DFB LDs. *Electron. Lett.*, 26:648–649, 1990.

[15] S. Norimatsu et al. PSK optical homodyne detection using external cavity laser diodes in costas loop. *IEEE Photonics Technol. Lett.*, 2(5):374–376, 1990.

[16] J. M. Kahn et al. Optical phaselocked receiver with multigigahertz signal bandwidth. *Electron. Lett.*, 25:626–628, 1989.

[17] J. M. Kahn. 1 Gb/s PSK homodyne transmission system using phase–locked semiconductor lasers. *IEEE Photonics Technol. Lett.*, 1(10):340, 1989.

[18] J. M. Kahn et al. 4 Gb/s PSK homodyne transmission system using phase–locked semiconductor lasers. *IEEE Photonics Technol. Lett.*, 2(4):285–287, 1990.

[19] D. A. Atlas and L. G. Kazovsky. An optical PSK homodyne transmission experiment using 1320 nm diode–pumped Nd:YAG lasers. *IEEE Photonics Technol. Lett.*, 2(5):367–370, 1990.

[20] J. M. Kahn. BPSK homodyne detection experiment using balanced optical phase locked loop with quantized feedback. *IEEE Photonics Technol. Lett.*, 2(11):840–842, 1990.

[21] L. G. Kazovsky and D. A. Atlas. A 1320 nm experimental optical phase–locked loop: performance investigation and PSK homodyne experiments at 140 Mb/s and 2 Gb/s. *J. Lightwave Technol.*, 8(9):1414–1425, 1990.

[22] G. Fischer et al. PSK optical homodyne systems operating near quantum limit. In *Proc. ECOC, European Conference on Optical Communications*, pages PDA-2, Gothenburg, Sweden, 1989.

[23] G. Fischer. A 700 Mb/s PSK optical homodyne system with balanced phase–locked loop. *J. Opt. Commun.*, 27–28, 1988.

[24] L. Kazovsky et al. A 1300 nm optical phase–locked loop. In *Proc. 15th European Conf. on Optical Communications*, pages 396–400, Gothenburg, 1989.

[25] T. G. Hodgkinson. Phase–locked loop analysis for pilot carrier coherent optical receivers. *Electron. Lett.*, 21:1202–1203, 1985.

[26] B. Wandernoth and J. Franz. Realization of a coherent optical DPSK heterodyne system with 565 Mbit/s at 1064 nm. In *Proc. SPIE: Optical Space Communication II*, pages 1522–1526, Munich, June vol. 1522, 1991.

Lecture Notes in Control and Information Sciences

Edited by M. Thoma and A. Wyner

Vol. 137: S. L. Shah, G. Dumont (Eds.)
Adaptive Control Strategies for
Industrial Use
Proceedings of a Workshop
Kananaskis, Canada, 1988
VI, 360 pages. 1989

Vol. 138: D. C. McFarlane, K. Glover
Robust Controller Design Using
Normalized Coprime Factor
Plant Descriptions
V, 206 pages. 1990

Vol. 139: V. Hayward, O. Khatib (Eds.)
Experimental Robotics I
The First International Symposium
Montreal, June 19-21, 1989
VII, 613 pages. 1990

Vol. 140: Z. Gajic, D. Petkovski,
X. Shen
Singularly Perturbed and Weakly
Coupled Linear Control Systems
A Recursive Approach
II, 202 pages. 1990

Vol. 141: S. Gutman
Root Clustering in Parameter Space
VIII, 153 pages. 1990

Vol. 142: A. N. Gündeş, C. A. Desoer
Algebraic Theory of Linear
Feedback Systems with Full and
Decentralized Compensators
V, 176 pages. 1990

Vol. 143: H.-J. Sebastian, K. Tammer (Eds.)
System Modelling and Optimization
Proceedings of the 14th IFIP Conference
Leipzig, GDR, July 3-7, 1989
X, 960 pages. 1990

Vol. 144: A. Bensoussan, J. L. Lions (Eds.)
Analysis and Optimization of Systems
Proceedings of the 9th International
Conference, Antibes, June 12–15, 1990
XII, 992 pages. 1990

Vol. 145: M. B. Subrahmanyam
Optimal Control with a Worst-Case
Performance Criterion and Applications
V, 133 pages. 1990

Vol. 146: D. Mustafa, K. Glover
Minimum Entropy H_∞ Control
X, 144 pages. 1990

Vol. 147: J. P. Zolésio (Ed.)
Stabilization of Flexible Structures
Third Working Conference,
Montpellier, France, January 1989
V, 327 pages, 1990

Vol. 148: In preparation

Vol. 149: K. H. Hoffmann, W. Krabs (Eds.)
Optimal Control of Partial
Differential Equations
Proceedings of the IFIP WG 7.2 International
Conference, Irsee, April 9-12, 1990
VI, 245 pages. 1991

Vol. 150: L. C. G. J. M. Habets
Robust Stabilization
in the Gap-topology
IX, 126 pages. 1991

Vol. 151: J. M. Skowronski,
H. Flashner, R. S. Guttalu (Eds.)
Mechanics and Control
Proceedings of the 3rd Workshop on
Control Mechanics
in Honor of the 65th Birthday of
George Leitmann
January 22-24, 1990, University of Southern
California
IV, 497 pages. 1991

Vol. 152: J. D. Aplevich
Implicit Linear Systems
XI, 176 pages. 1991

Vol. 153: O. Hájek
Control Theory in the Plane
X, 272 pages. 1991

Vol. 154: A. Kurzhanski, I. Lasiecka (Eds.)
Modelling and Inverse Problems of Control
for Distributed Parameter Systems
Proceedings of IFIP (W.G. 7.2)-IIASA Conference
Laxenburg, Austria, July 24-28, 1989
VII, 179 pages. 1991

Vol. 155: M. Bouvet, G. Bienvenu (Eds.)
High-Resolution Methods in Underwater
Acoustics
V, 249 pages. 1991

Lecture Notes in Control and Information Sciences

Edited by M. Thoma and A. Wyner

Vol. 156: R. P. Hämäläinen, H. K. Ehtamo (Eds.)
Differential Games –
Developments in Modelling and Computation
Proceedings of the Fourth International Symposium
on Differential Games and Applications
August 9-10, 1990, Helsinki University of Technology,
Finland
XIII, 292 pages. 1991

Vol. 157: R. P. Hämäläinen, H. K. Ehtamo (Eds.)
Dynamic Games in Economic Analysis
Proceedings of the Fourth International Symposium
on Differential Games and Applications
August 9-10, 1990, Helsinki University of Technology,
Finland
XIII, 311 pages. 1991

Vol. 158: K. Warwick, M. Kárný,
A. Halousková (Eds.)
Advanced Methods in Adaptive Control
for Industrial Applications
X, 331 pages. 1991

Vol. 159: X. Li, J. Yong (Eds.)
Control Theory of Distributed Parameter Systems
and Applications
Proceedings of the IFIP WG 7.2 Working Conference,
Shanghai, China, May 6–9, 1990
VIII, 219 pages. 1991

Vol. 160: P. V. Kokotović (Ed.)
Foundations of Adaptive Control
IX, 525 pages. 1991

Vol. 161: L. Gerencsér, P. E. Caines (Eds.)
Topics in Stochastic Systems:
Modelling, Estimation and Adaptive Control
IV, 401 pages, 1991

Vol. 162: C. Canudas de Wit (Ed.)
Advanced Robot Control
IX, 314 pages, 1991

Vol. 163: V. L. Mehrmann
The Autonomous Linear Quadratic
Control Problem
Theory and Numerical Solution
VI, 177 pages, 1991

Vol. 164: I. Lasiecka, R. Triggiani
Differential and Algebraic Riccati Equations
with Application to Boundary/Point
Control Problems: Continuous Theory
and Approximation Theory
XI, 160 pages, 1991

Vol. 165: G. Jacob, F. Lamnabhi-Lagarrigue (Eds.)
Algebraic Computing in Control
Proceedings of the First European Conference
Paris, March 13–15, 1991
IX, 385 pages. 1991

Vol. 166: L. L. M. van der Wegen
Local Disturbance Decoupling
with Stability for Nonlinear Systems
V, 135 pages, 1991

Vol. 167: M. Rao
Integrated System for Intelligent Control
VIII, 133 pages. 1992

Vol. 168: P. Dorato, L. Fortuna, G. Muscato
Robust Control for Unstructured Perturbation
An Introduction
VI, 117 pages. 1992

Vol. 169: V. M. Kuntzevich, M. Lychak
Guaranteed Estimates, Adaptation
and Robustness in Control Systems
IV, 209 pages. 1992

Vol. 170: J. M. Skowronski, H. Flashner,
R. S. Guttalu (Eds.)
Mechanics and Control
Proceedings of the 4th Workshop
on Control Mechanics, January 21–23, 1991
University of Southern California, USA
IV, 301 pages. 1992

Vol. 171: P. Stefanidis, A. P. Papliński,
M. J. Gibbard
Numerical Operations with Polynomial Matric
Application to Multi-Variable Dynamic
Compensator Design
VIII, 205 pages. 1992

Vol. 172: H. Tolle, E. Ersü
Neurocontrol
Learning Control Systems
Inspired by Neuronal Architectures
and Human Problem Solving Strategies
X, 211 pages. 1992

Vol. 173: W. Krabs
On Moment Theory and Controllability
of One-Dimensional Vibrating
Systems and Heating Processes
VII, 174 pages. 1992

Lecture Notes in Control and Information Sciences

Edited by M. Thoma and A. Wyner

Vol. 174: A.J.M. Beulens, H.-J. Sebastian (Eds.)
Optimization-Based Computer-Aided
Modelling and Design
Proceedings of the First Working Conference
of the IFIP TC 7.6 Working Group,
The Hague, The Netherlands, 1991
VIII, 270 pages, 1992

Vol. 175: E. Rogers, D.H. Owens
Stability Analysis for Linear Repetitive Processes
VII, 197 pages, 1992

Vol. 176: B.L. Rozovskii, R.B. Sowers (Eds.)
Stochastic Partial Differential Equations
and Their Applications
Proceedings of IFIP WG 7/1 International Conference
University of North Carolina at Charlotte, NC
June 6 - 8, 1991
VIII, 251 pages, 1992

Vol. 177: I. Karatzas, D. Ocone (Eds.)
Applied Stochastic Analysis
Proceedings of a US-French Workshop,
Rutgers University, New Brunswick, N.J.
April 29 - May 2, 1991
X, 311 pages, 1992

Vol. 178: J.P. Zolésio (Ed.)
Boundary Control and Boundary Variation
Proceedings of IFIP WG 7.2 Conference,
Sophia Antipolis, France
October 15 - 17, 1990
VIII, 392 pages 1992

Vol. 179: Z.H. Jiang, W. Schaufelberger
Block Pulse Functions and Their Applications
in Control Systems
XII, 237 pages, 1992

Vol. 180: P. Kall (Ed.)
System Modelling and Optimization
Proceedings of the 15th IFIP Conference
Zurich, Switzerland, September 2-6, 1991
XIX, 969 pages, 1992

Vol. 181: C.R. Drane
Positioning Systems - A Unified Approach
X, 168 pages 1992

Vol. 182: J. Hagenauer (Ed.)
Advanced Methods for Sattellite
and Deep Space Communications
Proceedings of an International Seminar
Organized by Deutsche Forschungsanstalt
für Luft- und Raumfahrt (DLR)
Bonn, Germany, September 1992
VII, 196 pages 1992